KEY MATHS 9³

▶ **David Baker**
The Anthony Gell School, Wirksworth

▶ **Paul Hogan**
Fulwood High School, Preston

▶ **Barbara Job**
Christleton County High School, Chester

▶ **Renie Verity**
Pensby High School for Girls, Heswall

Stanley Thornes (Publishers) Ltd

Contents

Acknowledgements

The publishers thank the following for permission to reproduce copyright material:
Allsport/Mike Cooper& Mike Hewitt: 57;
Janine Wiedel Photo: 85;
John Birdsall Photography: 67;
Martyn Chillmaid: 20, 25, 27, 33, 44, 54, 69, 102, 104, 116, 131, 142, 146, 227, 249, 328;
Rex Features: 116;
Science Photo Library: 93 (R Ressmeyer/Starlight);
Topham Picturepoint: 46;
TRH Pictures: 217;
All other photographs STP Archives

The publishers have made every effort to contact copyright holders but apologise if any have been overlooked.

First published in 1997 by
Stanley Thornes (Publishers) Ltd
Ellenborough House
Wellington Street
CHELTENHAM   GL50 1YW

00   01   /   10   9   8   7

A catalogue record for this book is available from the British Library.

ISBN 0 7487 2798 1

Original design concept by Studio Dorel
Cover design by John Christopher, Design Works
Cover photographs: Tony Stone Images (front);
Pictor International (spine); Tony Stone Images (back)
Artwork by Maltings Partnership, Hugh Neill,
David Oliver, Angela Lumley, Jean de Lemos
Cartoons by Clinton Banbury
Typeset by Tech Set Ltd
Printed and bound in China by Dah Hua Printing Press Co. Ltd.

Pythagoras

Pythagoras was a Greek philosopher and mathematician who lived in the sixth century BC.

Apart from his famous theorem, Pythagoras also discovered the mathematical basis of music. For example, that an octave can be expressed as the ratio 1:2. (A string stopped at half its length will sound the octave above the full length.)

1 Pythagorean triples

The Ancient Egyptians used ropes to make sure that their buildings had square corners.

Exercise 1:1

1 Look at this sequence of squares:

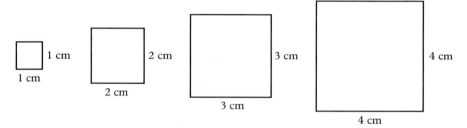

1 cm   2 cm   3 cm   4 cm

The areas of the squares are 1 × 1, 2 × 2, 3 × 3, 4 × 4
The areas of these squares give you the first four square numbers: 1, 4, 9, 16
Write down the first 16 square numbers.

2 Look at these square numbers:
  9   16   25
These three square numbers are special because
  9 + 16 = 25
There are other sets of square numbers which work like this.
a How many can you find?
b You can write
    9 + 16 = 25
  as $3^2 + 4^2 = 5^2$
When you do this the numbers 3, 4 and 5 are called a **Pythagorean triple**.
Use your answers to **a** to write your own Pythagorean triples.

3 Look at the Pythagorean triple 3, 4 and 5.

Here is a triangle with sides of
length 3 cm, 4 cm and 5 cm:

You are going to draw this triangle.

a Draw a line AB 5 cm long.

Open your compasses to 4 cm.
Put the point on A.
Draw an arc.

b Open your compasses to 3 cm.
Put the point on B.
Draw an arc.

c The two arcs cross at a point.
Mark this point C.
Draw in AC and BC.

d Measure the largest angle of your triangle.
Mark the size of this angle on your triangle.

4 Draw two more triangles using your triples from question 2.
Measure the largest angle of each triangle.
Mark the size of this angle on the triangle.

5 What do your three triangles have in common?

6 Draw three more right angled triangles.
Choose any lengths you like for the sides.
For each triangle:
a Measure the lengths of the sides.
b Square these numbers.
c Is the square of the largest number equal to the sum of the squares
of the other two numbers?

2 Finding the hypotenuse

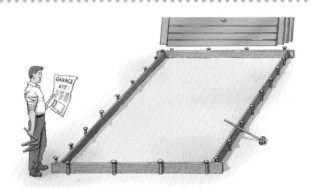

It is very important to have the right angles of the base correct before putting up the garage or it will not fit together properly. Mr Evans is going to measure the sides and a diagonal of this garage base.
He will use the lengths to make sure that he has a right angle in each corner.

Hypotenuse   The longest side of a right angled triangle is called the **hypotenuse**.

Pythagoras' theorem   In this right angled triangle:
$$c^2 = a^2 + b^2$$

The square of the = the sum of the squares of
    hypotenuse   the other two sides

This is called **Pythagoras' theorem** after the Greek mathematician, Pythagoras.

Example   **a** Write down the letter of the hypotenuse in each of these triangles.
            b Write down Pythagoras' theorem for each of these triangles.

(1)   (2)

a The hypotenuse is p **a** The hypotenuse is ST
b $p^2 = q^2 + r^2$ **b** $ST^2 = SR^2 + RT^2$

Exercise 1:2

a Write down the letter of the side that is the hypotenuse in each of these triangles.
b Write down Pythagoras' theorem for each of these triangles.

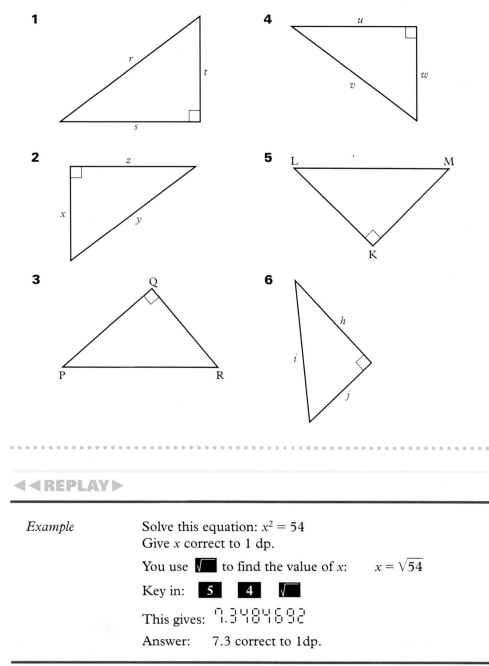

1

2

3

4

5

6

. . . .

◀◀REPLAY▶

Example

Solve this equation: $x^2 = 54$
Give x correct to 1 dp.

You use $\sqrt{}$ to find the value of x: $x = \sqrt{54}$

Key in: **5**   **4** $\sqrt{}$

This gives: 7.348469228

Answer: 7.3 correct to 1dp.

7 Solve these equations.

   a $x^2 = 81$ **b** $v^2 = 196$ **c** $p^2 = 361$ **d** $r^2 = 184.96$

8 Solve these equations.

Give your answers correct to 1 dp.

   a $n^2 = 35$ **c** $m^2 = 75$ **e** $q^2 = 80$ **g** $s^2 = 146$

   b $y^2 = 56.8$ **d** $x^2 = 45.1$ **f** $w^2 = 9.07$ **h** $k^2 = 166.3$

You can use Pythagoras' theorem to find the length of the hypotenuse if you know the lengths of the other two sides.

Example   Find the length of the hypotenuse of this triangle.

Give your answer correct to 1 dp.

Using Pythagoras' theorem:

$p^2 = 7^2 + 10^2$

$p^2 = 149$

$p = 12.2$ cm correct to 1 dp.

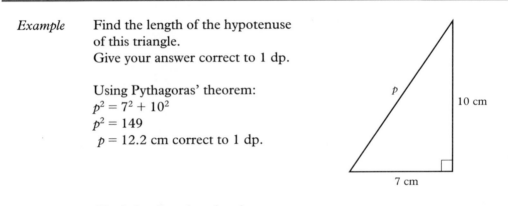

Check: Look at the triangle.

12.2 cm seems a reasonable length for the longest side.

Exercise 1:3

1 Find the length of the hypotenuse in each of these triangles.

Give your answers correct to 1 dp when you need to round.

Look at the triangle each time to check that your answer is reasonable.

a   **b**

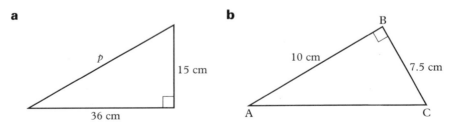

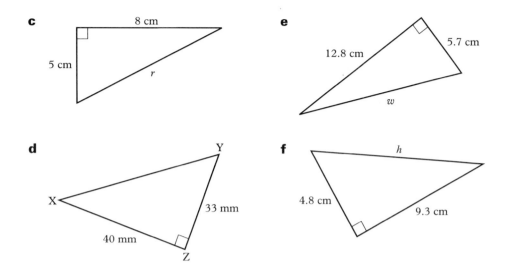

c 8 cm, 5 cm, r

e 12.8 cm, 5.7 cm, w

d Y, X, Z, 33 mm, 40 mm

f h, 4.8 cm, 9.3 cm

For the rest of this exercise, give your answers correct to 1 dp when you need to round.

2 Alan is orienteering.
He travels cross country as shown in the diagram.
He finishes 145 m north and 50 m west from his starting point.
How far is Alan from his starting point?

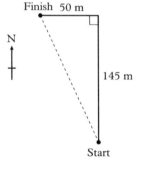

Finish 50 m

N

145 m

Start

3 Alisha is using a glass rod to stir the liquid in this beaker.
The rod is 25 cm long.
The beaker is 14 cm high and has a diameter of 10 cm.
How long is the part of the glass rod outside the beaker?

4 A mast is held in place by two equal wires as shown.
The height of the mast is 70 m.
Find the length of each wire.

50 m

5 This is Gavin's tent.
It is 80 cm high and 120 cm wide.
Find the length of one of the
sloping sides of the tent.

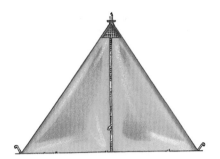

6 You cannot draw a right angled triangle with edges 7 cm, 9 cm and 21 cm.
Use Pythagoras' theorem to show this.

7 Mr Evans is making a base for a garage.
The sides of the base measure 3.5 m
and 5 m.
The base should be a rectangle.
Mr Evans finds a diagonal
measures 6.5 m.
Use Pythagoras' theorem to decide
if the base is a rectangle.

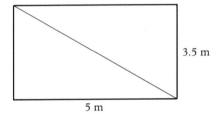

8 The diagram shows the
triangle EFG.
  a EG = 6 units
  Write down the length of FG.
  b Use Pythagoras' theorem to
  find the length of EF.

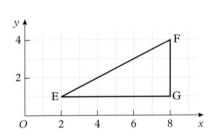

9 Draw a pair of axes on squared paper.
Plot points P (2, 1), Q (5, 8) and R (5, 1).
  a Find the length PR.
  b Find the length RQ.
  c Use Pythagoras' theorem to find the length PQ.

● 10 Find the distance between the points with co-ordinates:
  a (3, 5) and (4, 9)   **c** (−3, −5) and (−5, −9)
  b (−1, 0) and (2,5)   **d** (−8, 3) and (−2, −5)

3 Finding any side

Mark has to clean the gutter at the edge of the roof.
He wants to know if his ladder will reach the gutter.

Pythagoras' theorem can be used to find one of the shorter sides of a triangle.

Example   Mark's ladder is leaning against a wall.
Find the height h that the ladder
reaches up the wall.
Give your answer correct to 1 dp.

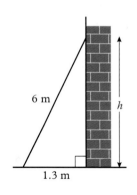

6 m

h

1.3 m

Using Pythagoras' theorem.

$$6^2 = h^2 + 1.3^2$$ Write the equation the other way round.

$$h^2 + 1.3^2 = 6^2$$ The unknown, h, is now on the left hand side.

$$h^2 + 1.3^2 - \mathbf{1.3^2} = 6^2 - \mathbf{1.3^2}$$ Subtract 1.3^2 from both sides.

$$h^2 = 34.31$$

$$h = 5.9 \text{ m correct to 1 dp}$$

Check: 5.9 m is less than the hypotenuse, 6 m. It looks like a reasonable answer.

Exercise 1:4

1   Find the missing lengths in these triangles.
    Give your answers correct to 1 dp when you need to round.

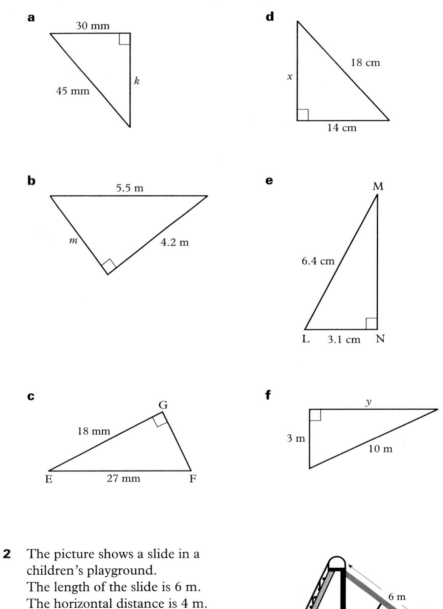

a 30 mm 45 mm k

d 18 cm x 14 cm

b 5.5 m m 4.2 m

e M 6.4 cm L 3.1 cm N

c G 18 mm E 27 mm F

f y 3 m 10 m

2   The picture shows a slide in a
    children's playground.
    The length of the slide is 6 m.
    The horizontal distance is 4 m.
    Find the height of the side.
    Give your answer correct to 2 dp.

6 m

4 m

3 This is a diagram of the water chute in a theme park.
The chute is 20 m long.
The horizontal distance travelled is 15.4 m.

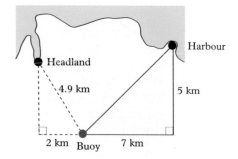

a Find the vertical distance travelled correct to 2 dp.

b Each step is 18 cm high.
How many steps will be needed to reach the top of the water chute?

4 Two sides of a right angled triangle are 14 cm and 18 cm.
How long will the third side be if
a it is the longest side   **b** it is the shortest side?
Give your answers correct to 1 dp.

5 A boat sails from the harbour to the buoy.
The buoy is 5 km south and 7 km west of the harbour.

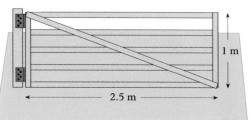

a How far has the boat travelled?

b The boat is now 4.9 km from the tip of the headland.
The tip is 2 km west of the boat.
How far north of the boat is it?
Give your answers correct to 2 dp.

· ·

Exercise 1:5

1 A gate needs to be strengthened by fixing a strut across the diagonal.
Calculate the length of the strut.
Give your answer correct to 2 dp.

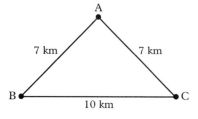

2 The positions of three houses are marked on the diagram.
House A is 7 km from B and 7 km from C.
The distance between B and C is 10 km along a straight road.
What is the shortest distance from A to the road?
Give your answer correct to 2 dp.

3 Give your answers correct to the nearest metre in this question.

Sonya is training for her Duke of Edinburgh Award. She is improving her fitness by running in the local park. She uses two different routes.

a Sonya's first route is to run the perimeter of the park.
How long is this route?

b Sonya's second route is shown by the arrows in the diagram.
How long is this route?

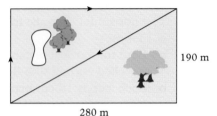

190 m

280 m

4 Rajiv is moving a fridge into his kitchen. He is using a trolley to wheel the fridge. The diagram shows the fridge when it is being moved by the trolley.

a Find the total height of the fridge plus the trolley correct to 2 dp.

b The kitchen door is 2 m high. Will Rajiv get the fridge through the door?

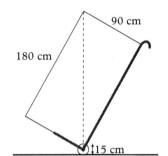

90 cm

180 cm

15 cm

5 The diagram shows the side of a ramp. Find the height of the ramp correct to 2 dp.

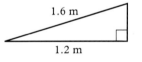

1.6 m

1.2 m

6 Triangle PQR is isosceles and has a right angle at Q.
The length of PR is 28 cm.
Find the length of side PQ correct to 1 dp.

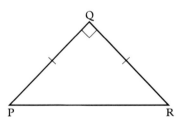

Q

P

R

7 Find the area of this equilateral triangle of side 26 cm.

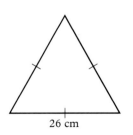

26 cm

1 One of these sets of three numbers is not a Pythagorean triple.
It does not obey the rule $c^2 = a^2 + b^2$
Find the odd one out.
a 10, 24, 26 **b** 8, 15, 17 **c** 4, 5, 6 **d** 6, 8, 10

2 An aircraft flies 260 km due north and then 340 km due east.
How far is it from its starting point?
Give your answer correct to the nearest kilometre.

3 This is a picture of a rubber.
Find the length of a sloping
edge of the rubber correct to
the nearest millimetre.

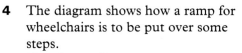

4 The diagram shows how a ramp for
wheelchairs is to be put over some
steps.
How long will the ramp need to be?
Give your answer correct to the
nearest centimetre.

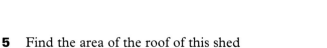

5 Find the area of the roof of this shed
correct to 1 dp.

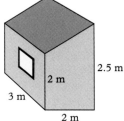

6 The end of this prism is an equilateral
triangle of side 8 cm.
  a Calculate the height of the triangle, h.
  b Find the area of the triangle.
  c Find the volume of the prism.
Give your answers correct to 1 dp.

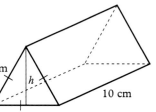

7 A cruise ship leaves the harbour and travels 35 km on a bearing of 050°. It then travels 84 km on a bearing of 140°.
It then turns and heads straight back to the harbour.
What is the total length of the cruise? correct to the nearest kilometre?

8 Jason has a theory that if he moves a ladder 1 m closer to a wall, it will reach 1 m further up the wall.
The diagrams show the two positions of Jason's ladder.
  a Find the distance up the wall that the ladder reaches in the two positions.
Give your answers correct to 2 dp.
  b How much further up the wall does the ladder reach in position (2)?
Is Jason's theory correct?

9 The triangle shown is isosceles.
  a Calculate the height of the triangle correct to 1 dp.
  b Work out the area of the triangle.

10 Jane is flying a kite.
Nathan is standing directly below the kite.
He is 14 m from Jane.
The kite string is 30 m long.
Work out the height of the kite.
Give your answer correct to the nearest metre.

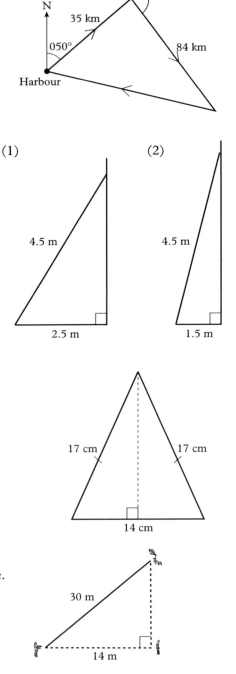

1 Each set of three numbers in the table obeys the rule $c^2 = a^2 + b^2$
There is another pattern in the table.

a	b	c	a^2
3	4	5	
5	12	13	
7	24	25	

a Copy the table.
Complete the a^2 column.
Write down what you notice.

b Add these rows to your table.
Use the pattern to fill in the gaps.

a	b	c	a^2
21	220		
25			

c Check that these values of a, b and c obey the rule $c^2 = a^2 + b^2$

2 Look at the spiral.
a Use Pythagoras' theorem to
find p^2.
b Use Pythagoras' theorem and
your value of p^2 to find q^2.
c Use the pattern to find the
length marked s.

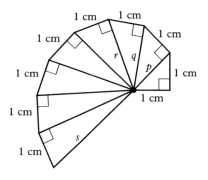

3 This cube has sides of length 5 cm.
Calculate the length of the internal
diagonal of the cube correct to
1 dp.
You will have to find another length
first.

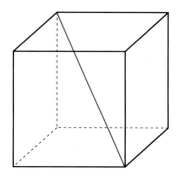

4 A room is 6 m long, 4 m wide and 3 m high.
It has two computer network points on opposite sides of the room as
shown in the diagram.
These network points have to be connected together using the shortest
possible length of cable.

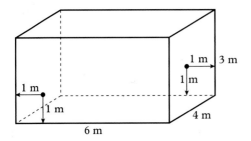

   a Sketch a suitable net of the room.
   b Mark the positions of the computer network points on the net.
   Join them with a line.
   Hence calculate the shortest possible length of cable needed.
   Give your answer correct to 2 dp.

5 The circle has a radius of 7 cm.
A chord of length 10 cm is
drawn in the circle.
Calculate the distance from the
centre of the circle to the chord below.
Give your answer correct to 1 dp.

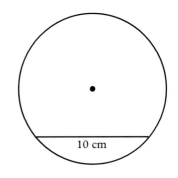

10 cm

6 **a** Use Pythagoras' theorem to write down an equation in x for each of
   these triangles.
   b Solve your equations (use trial and improvement in part (3)).
   Give your answers correct to 1 dp.

(1)   (2)   (3)

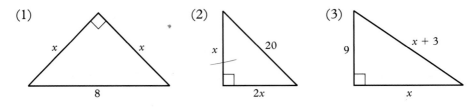

- **Hypotenuse**   The longest side of a right angled triangle is called the **hypotenuse**.

- **Pythagoras' theorem**   In this right angled triangle:
$$c^2 = a^2 + b^2$$

The square of the = the sum of the squares of
   hypotenuse   the other two sides

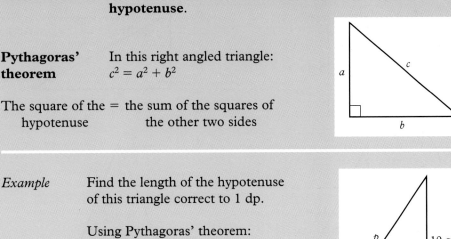

- *Example*   Find the length of the hypotenuse of this triangle correct to 1 dp.

Using Pythagoras' theorem:
$$p^2 = 7^2 + 10^2$$
$$p^2 = 149$$
$$p = 12.2 \text{ cm correct to 1 dp.}$$

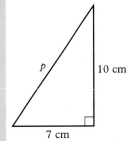

Check: Look at the triangle.
              12.2 cm seems a reasonable length for the longest side.

- *Example*   Mark's ladder is leaning against a wall.
Find the height *h* that the ladder reaches up the wall.
Give your answer correct to 1 dp.

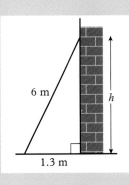

Using Pythagoras' theorem.

$$h^2 + 1.3^2 = 6^2$$
$$h^2 + 1.3^2 - \mathbf{1.3^2} = 6^2 - \mathbf{1.3^2} \qquad \text{Subtract } 1.3^2 \text{ from both sides.}$$
$$h^2 = 34.31$$
$$h = 5.9 \text{ m correct to 1 dp}$$

Check: 5.9 m is less than the hypotenuse, 6 m. It looks like a reasonable answer.

1 Calculate the lengths of the sides marked with letters in these triangles.
Give your answers correct to 1 dp when you need to round.

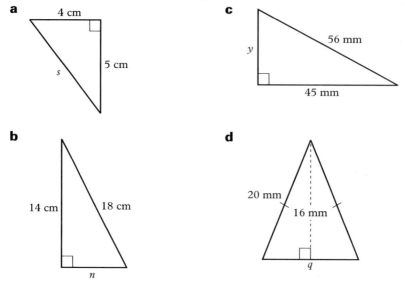

a 4 cm   5 cm   *s*

c *y*   56 mm   45 mm

b 14 cm   18 cm   *n*

d 20 mm   16 mm   *q*

2 Use Pythagoras' theorem to test whether this triangle has a right angle.

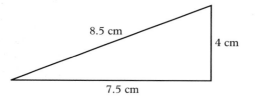

8.5 cm   4 cm   7.5 cm

3 A ladder 3.5 m long, leans against the wall of a house.
The foot of the ladder is 1.2 m from the wall.
Calculate how far the ladder reaches up the wall.
Give your answer correct to 2 dp.

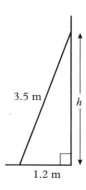

3.5 m   *h*   1.2 m

2 Formulas

For a moving object

$E = mc^2$

where E = energy,

c is the velocity of light

and m is mass

In simple terms this states that all energy has mass. It was devised by the German-born scientist Albert Einstein.

1 Formulas

Colin is going to paper his bedroom.
He needs to know how many rolls of wallpaper to buy.
Colin has found a formula for estimating the number of rolls of wallpaper:

$$R = \frac{H \times P}{5}$$

H = **H**eight of room in metres
P = **P**erimeter of room in metres

Colin's room is about 2 m high and the perimeter is 16 m.

Colin needs $\dfrac{2 \times 16}{5} = 6.4$ rolls

Colin decides to buy 7 rolls of paper.

◀◀**REPLAY**▶

Exercise 2:1

Here are some formulas used in maths for finding perimeters and areas.

1 $P = 4l$ gives the perimeter of a square.
   a Find P when $l = 5$ cm.
   b Find P when $l = 7$ cm.

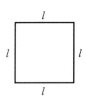

2 $A = l^2$ gives the area of a square.
   Remember: $l^2 = l \times l$
   a Find A when $l = 4$ cm.   **b** Find A when $l = 9$ cm.

3 $P = 2l + 2w$ gives the perimeter of a rectangle.
   a Find P when $l = 6$ cm and $w = 4$ cm.
   b Find P when $l = 8$ m and $w = 5$ m.

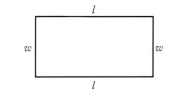

4 $A = lw$ gives the area of a rectangle.
   Find A when $l = 7$ cm and $w = 5$ cm.

5 $P = 2 \times (l + w)$ is another way of writing the perimeter of a rectangle.
Find P when $l = 4.5$ cm and $w = 2.5$ cm.

6 $A = 6l^2$ is a formula for finding the
surface area of a cube of side l.
Remember: $A = 6 \times l^2$
  a Find A when $l = 3$ cm.
  b Find A when $l = 5$ cm.

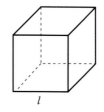

l

7 $A = \frac{1}{2}bh$ gives the area of a triangle.
Find A when $b = 30$ mm and $h = 20$ mm.

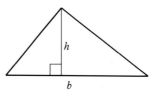

8 $A = bh$ gives the area of a parallelogram.
Find A when $b = 7$ cm and $h = 4$ cm.

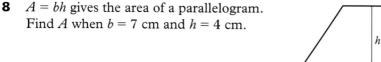

9 $C = 3d$ is a formula for estimating the
circumference of a circle.
  a Estimate C when $d = 15$ mm.
  b Estimate C when $d = 8$ cm.

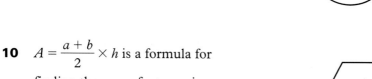

10 $A = \dfrac{a + b}{2} \times h$ is a formula for

finding the area of a trapezium.
Find A when $a = 5$ cm, $b = 7$ cm
and $h = 4$ cm.

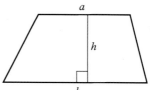

2

Some formulas are in two parts.
One part is an amount that varies, and the other part stays the same.

Example   The Green family hire a car for their holiday.
For each day of the holiday they must pay £50. They also pay a fixed amount of £40 for insurance.
a Write down a formula for the total cost (T = total, d = days).
b Use your formula to find T when $d = 12$ days.

a

Number of days d	Hire charge	Insurance	Total T
1	$50 \times 1 = 50$	40	$50 + 40 = 90$
2	$50 \times 2 = 100$	40	$100 + 40 = 140$
3	$50 \times 3 = 150$	40	$150 + 40 = 190$
4	$50 \times 4 = 200$	40	$200 + 40 = 240$

$T = 50 \times d + 40$ or $T = 50d + 40$

b When $d = 12$, $T = 50 \times 12 + 40 = 640$ Answer: £640

Exercise 2:2

1 The Jones family have one newspaper delivered to their house each day.
The newspaper costs 40 p. There is also a fixed delivery charge of 50 p per week.
a Copy the table and fill it in.

Number of days d	Cost of papers	Delivery charge	Total T
1	$40 \times 1 = 40$	50	$40 + 50 = 90$
2	$40 \times 2 = 80$	50	$80 + 50 = \ldots$
3			
4			

b Find a formula for the total cost of the papers.
(T = total cost in pence, d = number of days)
Use the table to help you.
c Use your formula to find T when (1) $d = 6$ days (2) $d = 4$ days

2 For each day's work in a shop, the manager is paid £60 and each of the assistants is paid £35.
a Find a formula for the total amount paid to the staff for one day.
(T = total, n = number of assistants)
b Use your formula to find T when (1) $n = 7$ (2) $n = 10$

3 Year 9 are having a party. It costs £90 to hire a disco and £3 per pupil for refreshments.
a Find a formula for the total cost of the party.
(T = total cost, n = number of pupils)
b Use your formula to find T when 120 Year 9 pupils go to the party.

22

4 The Northern Electricity Board charges customers 8 p per unit of electricity that they use, plus a fixed charge of £12.
    a Write down a formula for the total cost of electricity used.
       (T = total cost, n = number of units of electricity used)
    b The Patel family have used 1200 units of electricity.
       Use your formula to work out the total cost of their electricity bill.

5

A goods train has an engine 6 m long. Each wagon is 8 m long.
    a Write down a formula for the total length of the goods train.
       (T = total length, n = number of wagons)
    b Use your formula to find the total length of a train with 20 wagons.

6 Charlene has joined a swimming club. She had to pay £25 to join the club.
She also pays £1.50 every time she goes swimming.
    a Write down a formula for the total cost of Charlene's swimming sessions.
       (T = total cost, n = number of times Charlene goes swimming)
    b Use your formula to find the total cost if Charlene goes swimming 30 times.

- -

Exercise 2:3

Try each of the key sequences in this exercise yourself.
Use a scientific calculator.

1 Paul and Katie are working out $25 - 8 \times 2$
Paul does it in his head like this: $25 - 8 \times 2 = 17 \times 2 = 34$
Katie does the working on her scientific calculator:

      2 **5** **−** **8** **×** **2** **=**   Answer: 9

    a Which answer is correct?
    b What has the other person done wrong?

2 Gary and Sinita are working out $\dfrac{12 + 15}{3}$

Gary works the answer out on his scientific calculator:

$\boxed{1}\ \boxed{2}\ \boxed{+}\ \boxed{1}\ \boxed{5}\ \boxed{\div}\ \boxed{3}\ \boxed{=}$ Answer: 17

Sinita works the question out in her head like this: $\dfrac{12 + 15}{3} = \dfrac{27}{3} = 9$

a Which answer is correct?

b What has the other person done wrong?

3 Khalid and Sally have to work out $\dfrac{12}{2 \times 3}$

Khalid works it out in his head like this: $\dfrac{12}{2 \times 3} = \dfrac{12}{6} = 2$

Sally does the working on a calculator:

$\boxed{1}\ \boxed{2}\ \boxed{\div}\ \boxed{2}\ \boxed{\times}\ \boxed{3}\ \boxed{=}$ Answer: 18

a Which answer is correct?

b What has the other person done wrong?

You may need to use brackets before you can work an expression out.

Example

$m = \dfrac{I}{v - u}$ is a formula used in physics.

Use the formula to find m when $I = 30$, $v = 12$, $u = 8$

Write the value of each letter in the formula: $m = \dfrac{30}{12 - 8}$

Put brackets round the bottom: $m = \dfrac{30}{(12 - 8)}$

Key in: $\boxed{3}\ \boxed{0}\ \boxed{\div}\ \boxed{(}\ \boxed{1}\ \boxed{2}\ \boxed{-}\ \boxed{8}\ \boxed{)}\ \boxed{=}$

Answer: 7.5

---

4 In question **2** Gary worked out $\dfrac{12 + 15}{3}$ using his scientific calculator.
He got the answer wrong.

a Insert brackets in $\dfrac{12 + 15}{3}$ so that Gary can get the correct answer
using his scientific calculator.

b Check the answer using a scientific calculator.

c Do the same for $\dfrac{12}{2 \times 3}$ from question **3**.

Exercise 2:4

1 Work these out. Put brackets in where you need them.
Give your answers correct to 1 dp when you need to round.

a $\dfrac{54 + 42}{12}$ **c** $\dfrac{5.6 - 3.26}{2.5 + 2.3}$ **e** $\dfrac{314 - 142}{3.9 \times 9.6}$

b $\dfrac{81}{157 + 167}$ **d** $\dfrac{15^2}{12 \times 2.5}$ **f** $\dfrac{1}{1 - \sqrt{0.7}}$

2 The formula $w = \dfrac{P - 2l}{2}$ gives the
width of a rectangle. P is the
perimeter and l is the length.
Find w for this rectangle.

perimeter = 42 cm

length = 14.8 cm

3 The formula $6r(r + h)$ gives an estimate
for the *total* surface area of a cylinder.
h is the height of the cylinder and r is the
radius.
Use the formula to estimate the area of
aluminium sheet used to make this can.
The height of the can is 11.5 cm, and the
radius is 3.8 cm.

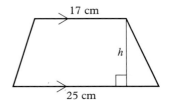

4 $h = \dfrac{2A}{a + b}$ gives the height of a trapezium.

A is the area and a and b are the parallel sides.
Find h for this trapezium.
The area of this trapezium is 273 cm².

17 cm

h

25 cm

5 ABCD is called a golden rectangle.
It is a shape that artists like.
The length of a golden rectangle
is found by multiplying the width

by $\dfrac{2}{\sqrt{5} - 1}$

Find the length of the golden
rectangle that has a width of 7 cm.

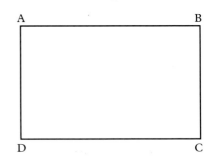

A   B

D   C

It is helpful to use brackets if you have more than one term under a square root.

Example

Work out $7\sqrt{4^2 + 3^2}$

Put brackets under the square root: $7\sqrt{4^2 + 3^2} = 7\sqrt{(4^2 + 3^2)}$

You start by working out the brackets:

| (| 4 | x^2 | + | 3 | x^2 |) | √ | × | 7 | = |

Answer: 35

Exercise 2:5

1 Work through the example. Make sure you know how to get the correct answer. Your calculator may need a different key sequence.

2 Work these out.
Give your answers correct to 1 dp when you need to round.

  a $2\sqrt{6^2 - 3^2}$ **b** $\frac{1}{2}\sqrt{17^2 - 12^2}$ **c** $1 + \sqrt{4.8^2 + 3.6^2}$

3 Theo hits a ball straight up in the air at a speed u of 15 m/s.
After 1 second the height of the ball, h, is 10 m.
Theo has a formula for finding the velocity v of the ball:
$v = \sqrt{u^2 - 20h}$
Use Theo's formula to find v.

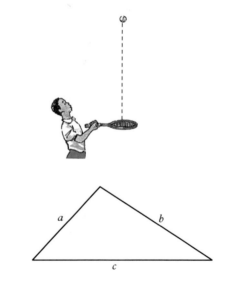

4 $\sqrt{s(s - a)(s - b)(s - c)}$ is a formula for finding the area of a triangle from the lengths of its sides.

$s = \dfrac{a + b + c}{2}$ and a, b, c are the

lengths of the three sides.
Find the area of this triangle.
$a = 6$ cm, $b = 7$ cm, $c = 10$ cm
Give your answer correct to 2 dp.

2   Patterns

Jane is making patterns using matchsticks.

◄◄REPLAY►

Exercise 2:6

1   Here are some of Jane's matchstick patterns.

Pattern 1   Pattern 2   Pattern 3

a   Sketch the next two patterns.
b   Copy this table and fill it in.

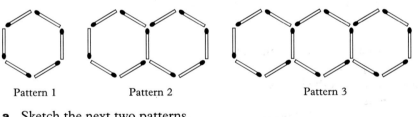

Number of pattern	1	2	3	4	5
Number of matchsticks	6	11			

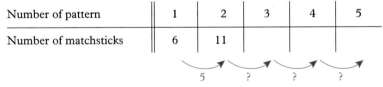

c   How many matchsticks do you add each time?
d   These are the first five multiples of 5:   5, 10, 15, 20, 25
        What do you add to these numbers to get the numbers of
        matchsticks?
e   Copy and complete:
        number of matchsticks = … × number of the pattern + …
f   Write down the formula using algebra. Use *m* for the number of
        matchsticks and *n* for the *number* of the pattern.

2 These patterns are made from yellow and blue counters.

Pattern number 1   Pattern number 2   Pattern number 3

a Draw the next two patterns.
b Copy this table and fill it in.

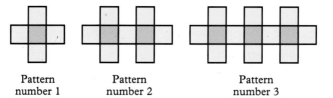

Number of blue counters	1	2	3	4	5
Number of yellow counters	4	6			

c How many yellow counters do you add each time?
d These are the first five multiples of 2: 2, 4, 6, 8, 10.
What do you add to these numbers to get the numbers of yellow counters?
e Copy this and fill in the formula for the number of yellow counters.
   number of yellow counters = ... × number of blue counters + ...
f Write down the formula using algebra.
Use y for the number of *yellow* counters and b for the number of *blue* counters.

3 These patterns are made of blue and yellow tiles.

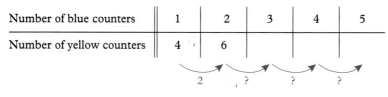

Pattern   Pattern   Pattern
number 1   number 2   number 3

a Draw the next two patterns.
b Make a table for the patterns showing the number of blue tiles and the number of yellow tiles.
c How many yellow tiles do you add each time?
d How many yellow tiles are there in the 10th pattern?
e Write down a formula in words for finding the numbers of yellow tiles from the number of blue tiles.
f Write down the formula using algebra. Use y for the number of *yellow* tiles and b for the number of *blue* tiles.

Exercise 2:7

1 Here is a sequence of squares.
1 cm is added to the sides of the squares each time.

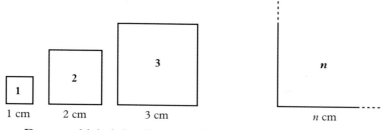

a Draw and label the diagrams for squares 4 and 5 of the sequence.
b Copy this table.
    Fill it in for the perimeters of the first 5 squares.

Number of square	1	2	3	4	5	n
Perimeter in cm	4×1	4×2	4×3	... × ...	... × ...	?

c Describe the number pattern shown in the table in words.
d Fill in the nth term using algebra.
e The areas of the squares are $1 \times 1 = 1^2$, $2 \times 2 = 2^2$, $3 \times 3 = 3^2$ etc.
    Copy this table and fill it in for the areas of the squares.

Number of square	1	2	3	4	5	n
Area in cm^2	1^2	2^2				?

2 Here is a sequence of cubes.

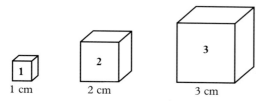

The volumes of the cubes are $1 \times 1 \times 1 = 1^3$ etc.

a Draw and label the diagrams for cubes 4 and 5 of the sequence.
b Copy the table and fill it in for the first 5 cubes.

Number of cube	1	2	3	4	5	n
Volume of cube in cm^3	1^3					

c Describe the number pattern shown in the table in words.
d Fill in the nth term using algebra.

The first three terms are given for each of the sequences in questions **3** to **5**.
a Draw and label shapes 4 and 5 of each sequence.
b Copy the table and fill it in for the first 5 shapes.
c Describe each number pattern in words.
d Fill in the nth term for each number pattern using algebra.

3 Here is a sequence of equilateral triangles.

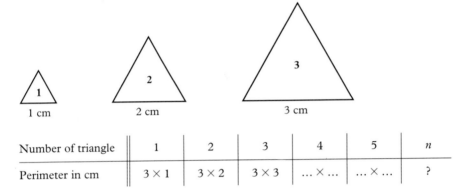

Number of triangle	1	2	3	4	5	n
Perimeter in cm	3×1	3×2	3×3	$... \times ...$	$... \times ...$	?

4 Here is a sequence of rectangles.

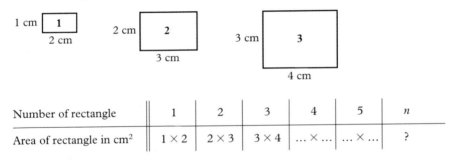

Number of rectangle	1	2	3	4	5	n
Area of rectangle in cm²	1×2	2×3	3×4	$... \times ...$	$... \times ...$	?

5 Here is a different sequence of rectangles.

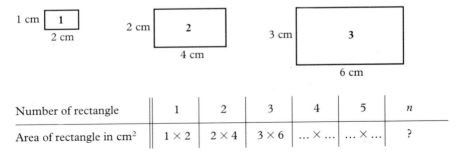

Number of rectangle	1	2	3	4	5	n
Area of rectangle in cm²	1×2	2×4	3×6	$... \times ...$	$... \times ...$	?

Formulas that contain n²

Example

Find a formula for the *n*th term of this number sequence:
6, 15, 28, 45, 66, ...

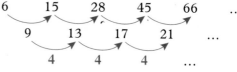

First differences are not the same

Second differences *are* the same

Formulas that contain n^2

If the **second difference is a constant**, the formula for the **nth term contains n^2**.
The number in front of n^2 is **half** the constant difference.

In the example, the constant difference is 4.
The number in front of n^2 is half of 4, which is 2.
The first part of the formula is therefore $2n^2$.

You need to make a table to help you find the rest of the formula.

Line 1: the number of each term
Line 2: the numbers in the sequence
Line 3: $2n^2$ worked out from the number of each term
Line 4: the 'Rest of sequence' comes from taking $2n^2$ away from the numbers in the sequence (Line 2 take away Line 3)

1. Number of term (*n*)	1	2	3	4	5
2. Sequence	6	15	28	45	66
3. $2n^2$	2	8	18	32	50
4. Rest of sequence	4	7	10	13	16

3 3 3 3

The formula for the 'Rest of sequence' is $3n + 1$
(since it is the multiples of 3, plus 1).

The final formula is $2n^2 + 3n + 1$

Check the formula by finding term number 6.
$n = 6$, $2n^2 + 3n + 1 = 91$
91 should fit the difference pattern at the start of the example.

Exercise 2:8

For the number sequences in questions **1** to **8**:
a Find the formula for the *n*th term.
b Find the 6th term using your formula and check to see if it fits the sequence.

1 3, 7, 13, 21, 31, ...

2 6, 11, 18, 27, 38, ...

3 3, 10, 21, 36, 55, ...

4 6, 15, 28, 45, 66, ...

5 9, 20, 37, 60, 89, ...

6 2, 4, 7, 11, 16, ...

7 3, 11, 23, 39, 59, ...

● **8** 0, 11, 28, 51, 80, ...

Solar panels

This space station is made of lots of small cubes fixed together. It has $4 \times 4 \times 4 = 64$ small cubes altogether.

The small cubes on the outside of the space station have circular solar panels. There is one solar panel fitted on every outside face.

1 a How many small cubes have 3 solar panels?
b How many small cubes have exactly 2 solar panels?
c How many small cubes have exactly 1 solar panel?
d How many small cubes have no solar panels?

2 Another space station is a cube with edges 3 small cubes long.
a Sketch this space station.
b How many small cubes of this space station have 3 solar panels?
c How many small cubes have exactly 2 solar panels?
d How many small cubes have exactly 1 solar panel?
e How many small cubes have no solar panels?

3 Investigate the numbers of solar panels in other sizes of space station.

3 Trial and improvement

Paul is making a curry.
He is adding the curry powder.
Paul wants the curry not to be too
mild nor too spicy.
He can taste the curry to see if it
is hot enough.
He can add more curry powder if
necessary.
If it is too spicy he can make it
milder next time.

You can solve equations by trying different values to get closer to the
right answer.
This is called trial and improvement.

◀◀REPLAY▶

Example   Solve $x^2 + 3x = 82$
Use trial and improvement.
Give your answer correct to 1 dp.

Value of x	Value of $x^2 + 3x$	
8	88	too big
7.5	78.75	too small
7.6	80.56	too small
7.7	82.39	too big x lies between 7.6 and 7.7

(Try 7.65 to see if x is between 7.6 and 7.65, or between 7.65 and 7.7)

7.65	81.4725	too small

x must be between 7.65 and 7.7
x rounds to 7.7
Answer: $x = 7.7$ to 1 dp

Exercise 2:9

Solve the equations in questions **1** to **7** by trial and improvement.
Give each answer correct to 1 dp.

1 $x^2 + x = 53$ **4** $x^2 + 3x = 145$

2 $x^2 + 2x = 82$ **5** $115 = x^2 + 2x$

3 $x^2 + x = 134$ **6** $71 = x^2 + 4x$

7 $x(x + 1) = 48$ *Remember:* $x(x + 1) = x \times (x + 1)$

8 The value of x lies between a pair of 1 decimal place numbers.
Find these numbers using trial and improvement.
a $x^2 + 3x = 13$ **b** $x^2 + x = 28$

9 Keith is investigating rectangles that have a perimeter of 20 cm.
   a Keith draws one of the rectangles.
   It has a width of 3 cm.
   (1) What is its length?
   (2) What is its area?

   3 cm

   ?

   b Copy this table to find more of Keith's rectangles.
   Fill in the table.

Width (cm)	Length (cm)	Area (cm²)
1		
2		
3	7	21
4		
5		

   c What does the width plus the length add up to each time?

   Keith wants a rectangle with perimeter 20 cm and area 22 cm².
   Keith calls the width of his rectangle x. The length is $10 - x$.
   Keith writes the equation $x(10 - x) = 22$

   d Find x using trial and improvement.
   Give the value of x correct to 1 dp.

Value of x	Value of $10 - x$	Area
3	7	21

   e Write down the lengths of the two sides of the rectangle for this
   value of x.

You can solve equations using trial and improvement correct to as many decimal places as necessary.

Example $x^2 + x = 49$ x lies between 6 and 7.
Give your answer correct to 2 dp.
Make a table and use trial and improvement.

Value of x	Value of $x^2 + 3x$	
7	56	too big
6.5	48.75	too small
6.6	50.16	too big x lies between 6.5 and 6.6
6.55	49.4525	too big
6.52	49.0304	too big
6.51	48.8901	too small x lies between 6.51 and 6.52
(Try 6.515 to see if x is between 6.51 and 6.515, or between 6.515 and 6.52)		
6.515	48.9602 ...	too small

x lies between 6.515 and 6.52
x rounds to 6.52
Answer: $x = 6.52$ to 2 dp

Exercise 2:10

For the equations in questions **1** to **6**:
Make a table and use trial and improvement.
Give each answer correct to 2 dp.

1 $x^2 + x = 79$

2 $x^2 + 2x = 19$

3 $x^2 + 4x = 93$

4 $39 = x^2 + 3x$

5 $x(7 + x) = 11$

6 $x(24 + x) = 110$

● **7**   Look at this equation: $x^2 + x = 65$
Make a table and use trial and improvement.
Give your answer correct to 3 dp.

1 Work these out.
Give your answer correct to 2 dp when you need to round.

a $\dfrac{6.6 + 8.5}{9.2 - 3.9}$ **b** $\dfrac{1.8^2 + 2.4^2}{0.85}$ **c** $\frac{1}{2}\sqrt{5.2^2 - 2.5^2}$ **d** $\dfrac{10}{\sqrt{14^2 - 9^2}}$

2 The area of this triangle can be found using the formula
Area $= \frac{1}{2}b\sqrt{a^2 - b^2}$

    a Calculate the area when
       $a = 5$ cm and $b = 1.4$ cm
    b Calculate the area when
       $a = 4.6$ cm and $b = 3.7$ cm
       Give your answers correct to 1 dp.

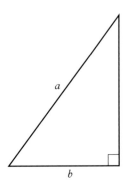

3 For each of these number sequences:
    a Find the formula for the nth term.
    b Find the 6th term using your formula and check to see if it fits the sequence.
       (1)   14, 18, 22, 26, 30, ...   (5)   7, 18, 29, 40, 51, ...
       (2)   7, 11, 17, 25, 35, ...   (6)   11, 23, 43, 71, 107, ...
       (3)   5, 14, 27, 44, 65, ...   (7)   4, 15, 32, 55, 84, ...
       (4)   7, 9, 12, 16, 21, ...   (8)   9, 24, 49, 84, 129, ...

4 Rani is making these patterns using red and yellow tiles.

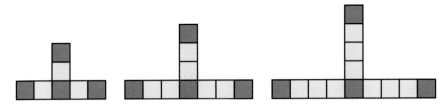

    a How many tiles will there be altogether in pattern number 4?
    b How many tiles will there be altogether in pattern number 8?
    c Rani's rule for the pattern is $3n + 4$.
       What does the 4 stand for in the $3n + 4$ rule?
    d What does the $3n$ stand for in the $3n + 4$ rule?
    e Rani wants to make pattern number 11.
       How many yellow tiles does she need?
    f How many tiles altogether does Rani need to make pattern number 11?
    g Rani makes a pattern using 34 tiles altogether.
       How many yellow tiles does she use?

5 The $\sqrt{}$ key on Keith's calculator is broken.

Keith is using trial and improvement to find $\sqrt{11}$

Here is Keith's working:

1st try $3 \times 3 = 9$
2nd try $3.4 \times 3.4 = 11.56$
3rd try $3.3 \times 3.3 = 10.89$

a Continue Keith's trials.
Give at least 4 more sensible trials.
Try to get as close to 11 as you can.
b Solve the equation $x^2 - 3 = 8$
You can use your working in part **a** to help you.

6 Lesley wants to find the values of x which make the equation
$3x^2 = 10x - 4$ correct.
Lesley works out the values of $3x^2$ and $10x - 4$.
She subtracts the value of $10x - 4$ from the value of $3x^2$.
This gives Lesley the *difference* between each pair of values.
Lesley writes down whether the difference is positive or negative.

x	$3x^2$	$10x - 4$	Difference	
-2	12	-24	36	positive
-1	3	-14	16	positive
0	0	-4	4	positive
1	3	6	-3	negative
2	12	16	-4	negative

a Lesley knows there is a value of x between 0 and 1 which makes the equation correct.
Use Lesley's table to explain why.
b Lesley tries some 1 decimal place numbers for x.

x	$3x^2$	$10x - 4$	Difference	
0.1	0.03	-3	3.03	positive
0.2	0.12	-2	2.12	positive

Copy Lesley's table.
The value of x lies between a pair of 1 decimal place numbers.
Continue Lesley's table to find these numbers.

1 $E = \frac{1}{2}mv^2$ is a formula used in maths and science.
Find the value of E for these values of m and v:
    a $m = 0.6, v = 7$ **b** $m = 2.5, v = -9$ **c** $m = 6, v = -15$

2   Here are some triangle numbers.

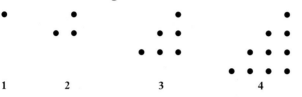

     1   2   3   4

    a   Draw diagrams for the fifth and sixth triangle numbers.
    b   Find a formula for the nth term of the triangle numbers.
    c   Use your formula to work out the 31st triangle number.
    d   Use trial and improvement to find the first triangle number over
        1000.

3   **a**   Copy this table for the equation $y = x^2$. Fill it in.

x	-3	-2	-1	-0.5	0	0.5	1	2	3
y	9							4	

    b   Use graph paper.
        Draw an x axis from -3 to 3 and a y axis from -4 to 10.
    c   Plot the points in your table for $y = x^2$.
        Join the points with a smooth curve.
    d   Label your graph $y = x^2$.
    e   Copy this table for the equation $y = 2x + 2$ and fill it in.

x	1	2	3
y	3		

    f   Plot the points and join them with a ruler to get the line $y = 2x + 2$.
    g   The line $y = 2x + 2$ should cut the curve $y = x^2$ in two points.
        Write down the x co-ordinate of each of the two points as
        accurately as you can. These values of x are the roots of the
        equation $x^2 = 2x + 2$. The roots are the values of x that make the
        equation correct.
    h   Use trial and improvement to find each of the values of x to 3 dp.
        The equation can be rearranged first like this:
$$x^2 = 2x + 2$$
$$x^2 - 2x = 2x + 2 - 2x$$
$$x^2 - 2x = 2$$

- You may need to use brackets before you can work an expression out.

  Example   Work out $\dfrac{12 + 15}{3}$

  Put brackets round the top. $\quad \dfrac{12 + 15}{3} = \dfrac{(12 + 15)}{3}$

  | (| 1 | 2 | + | 1 | 5 |) | ÷ | 3 | = |

  Answer: 9

- It helps to use brackets if you have more than one term under a square root.

  Example   Work out $7\sqrt{4^2 + 3^2}$

  Put brackets under the square root: $\quad 7\sqrt{4^2 + 3^2} = 7\sqrt{(4^2 + 3^2)}$
  You start by working out the brackets:

  | (| 4 | x^2 | + | 3 | x^2 |) | $\sqrt{}$ | × | 7 | = |

  Answer: 35

- **Formulas that contain n^2**   If the **second difference is a constant**, the formula for the **nth term contains n^2**.
  The number in front of n^2 is **half** the constant difference.

- You can solve equations by trying different values to get close to the right answer.

  Example $x^2 + x = 49$ x lies between 6 and 7.
  Give your answer correct to 2 dp.
  Make a table and use trial and improvement.

Value of x	Value of $x^2 + 3x$	
7	56	too big
6.5	48.75	too small
6.6	50.16	too big x lies between 6.5 and 6.6
6.55	49.4525	too big
6.52	49.0304	too big
6.51	48.8901	too small x lies between 6.51 and 6.52

  (Try 6.515 to see if x is between 6.51 and 6.515, or between 6.515 and 6.52)

6.515	48.9602 ...	too small

  x lies between 6.515 and 6.52
  x rounds to 6.52
  Answer: $x = 6.52$ to 2 dp

1 Some of Year 9 go on an outing to a castle. It costs £70 for the hire of the coach and £2 per pupil to go in the castle.
  a Write down a formula for the total cost of the outing.
    (T = total cost, n = number of pupils on the outing)
  b Use your formula to find the total cost of an outing for 54 pupils.

2 Find the value of D in each of these formulas.

  a $D = \dfrac{e^2 + f^2}{e - f}$ $e = 14.8,\ f = 12.3$

  b $D = 3\sqrt{g^2 + h^2}$ $g = 19,\ h = 14$

  Give your answers correct to 1 dp.

3 These patterns are made of square tiles.

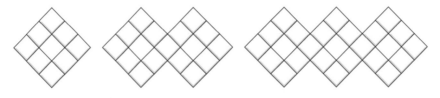

Pattern number 1   Pattern number 2   Pattern number 3

  a Draw the next two patterns.
  b Make a table for the patterns.
  c How many tiles do you add each time?
  d How many tiles are there in the 25th pattern?
  e Write down a formula for the number of tiles in the nth pattern.

4 For each of these number sequences:
  a Find the formula for the nth term.
  b Find the 6th term using your formula and check it using the sequence.
    (1) 8, 15, 22, 29, 36, ...   (2) 9, 19, 33, 51, 73, ...

5 Solve the equation $x^2 + 2x = 46$ using trial and improvement.
  Give the answer correct to 2 dp.

6 In the equation $x(14 + x) = 45$ the value of x lies between a pair of 1 decimal place numbers. Find x using trial and improvement.

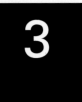

3 Circles

Here is a photograph taken in the Palais de la découverte, a science museum in Paris. The photograph shows the value of π, written to 706 decimal places, around the walls of the museum. Try to find out what the world record for calculating π is.

1   How long is a circle?

Kiran has just bought a new bike.
He wants to set up the computer
to be able to see how far he rides.
He enters the distance across the
wheel into the computer.
This is because the distance
round a circle depends on the
distance across the circle.

Exercise 3:1

W 1   You need worksheets 3 : 1 and 3 : 2.

2   The distance across each circle is 5 cm.
    a   Divide each of your estimates by 5.
       Use a calculator and write down all the numbers on the calculator
       display.
    b   What do you notice about these answers?

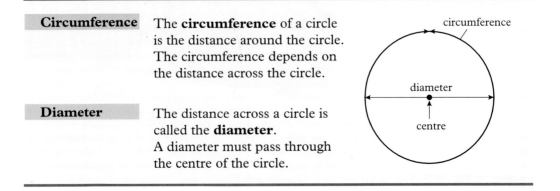

Circumference   The **circumference** of a circle
is the distance around the circle.
The circumference depends on
the distance across the circle.

Diameter   The distance across a circle is
called the **diameter**.
A diameter must pass through
the centre of the circle.

The circumference divided by the diameter always gives the same
answer.
This answer is a special number in maths.
It is called **pi** (which you say as 'pie') and it is written as π.

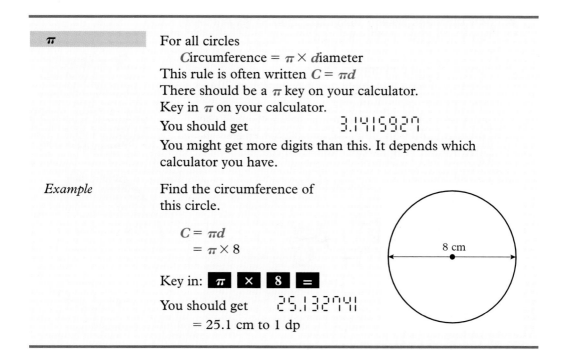

For all circles
    Circumference = $\pi \times$ diameter
This rule is often written $C = \pi d$
There should be a π key on your calculator.
Key in π on your calculator.
You should get   3.1415929
You might get more digits than this. It depends which
calculator you have.

Example   Find the circumference of
this circle.

$$C = \pi d$$
$$= \pi \times 8$$

Key in: $\boxed{\pi}$ $\boxed{\times}$ $\boxed{8}$ $\boxed{=}$
You should get   25.132741
        $= 25.1$ cm to 1 dp

8 cm

Exercise 3:2

Give all your answers in this exercise correct to 1 dp.

1   Find the circumferences of these circles.

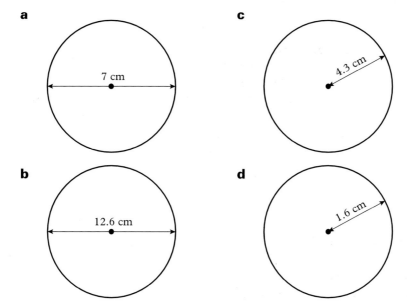

a

7 cm

c

4.3 cm

b

12.6 cm

d

1.6 cm

2 The diameter of each wheel on Kiran's bike is 70 cm.
What is the circumference of a wheel?

3 Janet is making a lampshade.
The top of the lampshade has a diameter of 20 cm.
The bottom has a diameter of 32 cm.
Janet wants to decorate the top and bottom of the
lampshade by putting braid around the edges.
What length of braid does she need to buy?

4 This circular mirror is edged with
decorative plastic tape.
What is the length of the tape?

8.5 cm

5 This plate is edged with gold trim.
What is the length of the gold trim?

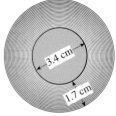

2.7 cm

6 The diagram shows a roll of sticky tape.
The diameter of the cardboard ring is 3.4 cm.
The tape on the roll is 1.7 cm thick.
What is the outer circumference of the tape?

3.4 cm

1.7 cm

7 The regions on this dartboard are separated using
lengths of wire.
There are 6 wire circles.
The diameters of these are 33.6 cm, 31.6 cm,
21 cm, 19 cm, 2.4 cm and 1.3 cm.
a What is the length of each straight piece of wire?
b What is the total length of all the straight pieces?
c Find the length of each circular wire.
d What is the total length of all the circular
pieces of wire?
e What is the total length of wire on the dartboard?

John is making a trundle wheel for his
Technology project.
He wants each turn of the wheel to
measure 1 metre on the ground.
The circumference of the wheel must
be 1 metre.
He needs to know what diameter to
use when he makes the wheel.
He needs a formula for the diameter.

The formula for finding the circumference is

$$C = \pi \times d$$

The inverse of multiplying by π is dividing by π.
Divide both sides by π:

$$\frac{C}{\pi} = \frac{\pi \times d}{\pi}$$

$$\frac{C}{\pi} = d$$

The formula for finding the diameter is

$$d = \frac{C}{\pi}$$

Example

The circumference of a circle is 40 cm.
Find the diameter.

$$d = \frac{C}{\pi}$$

$$d = \frac{40}{\pi}$$

Key in: 4 0 ÷ π = 12.732395

= 12.7 cm to 1 dp

---

8 Find the diameter that John must use to make his trundle wheel the
right size.
Give your answer in centimetres.

9 A roundabout has a circumference of 220 m.
What is the diameter of the roundabout?

10 What is the diameter of a circular mirror with circumference 250 cm?

Radius	The **radius** of a circle is the distance from the centre to the circumference.
	The radius is half the diameter.

Exercise 3:3

Give all your answers in this exercise correct to 1 dp.
For each question estimate the answer using $\pi = 3$.

1 For each of these circles find:
   a the diameter   **b** the radius

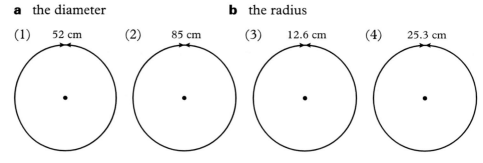

(1)   52 cm   (2)   85 cm   (3)   12.6 cm   (4)   25.3 cm

2 The largest big wheel in the world is in Kobe in Japan.
The circumference is 200 m.
Find the radius of the wheel.

3 The distance around a circular running track is 400 m.
Find the radius of the track.

4 The perimeter of this track is 400 m.
What is the radius of the semi-circular ends?

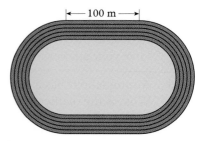

100 m

2 Area of a circle

Fred the farmer is annoyed about this crop circle.
His crop has been destroyed.
He wants to know how much of the crop has been lost.
He needs to know the area of the circle.

Area of a circle   The **area of a circle** depends on the radius of the circle. The formula is

Area of circle $= \pi \times$ radius $\times$ radius

This rule is often written $A = \pi \times r \times r$

or $\qquad A = \pi r^2$ $\qquad$ (r^2 means $r \times r$)

Example   Find the area of this circle.

$A = \pi r^2$
$\quad = \pi \times 4^2$
$\quad = \pi \times 4 \times 4$
$\quad = \pi \times 16$
$\quad = 50.3$ cm to 1 dp

4 cm

Exercise 3:4

Give all your answers in this exercise correct to 1 dp.

1 Find the areas of these circles.

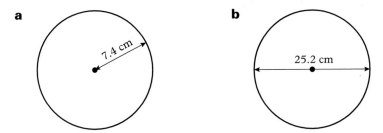

a 7.4 cm

b 25.2 cm

2 The diameter of the centre circle on a soccer pitch is 10 yds.
What is the area of the circle?

3 A CD is made from plastic.
The diameter of a CD is 12 cm.
The hole in the middle is 1.5 cm wide.
a Find the area of the hole.
b Find the area of the plastic.

4 Use $\pi = 3$ to *estimate* the areas of the circles in question **1–3**.

5 The floor of a disco is a circle with radius 4.5 m.
The floor needs polishing.
a What is the area of the floor?
b Polishing costs £2.50 for every square metre.
Any extra part of a square metre also costs £2.50.
How much does it cost to have the floor polished?

6 This circular pond has a diameter of 25 m.
There are 3 circular islands in the pond.
Each island has a radius of 2.5 m.
Find the area of the surface of the water.

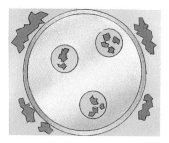

7 The diagram shows the cross section of a hose pipe.
The internal diameter is 1.2 cm.
The external diameter is 1.6 cm.
Find the area of the cross section
a in cm^2
b in mm^2

Exercise 3:5

Give all your answers in this exercise correct to 1 dp.

1 Find the areas of these shapes.

a

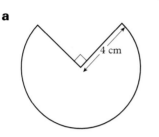

4 cm

c

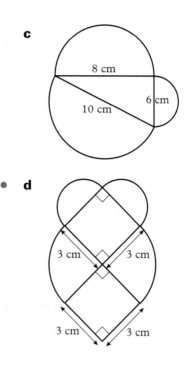

8 cm

6 cm

10 cm

b

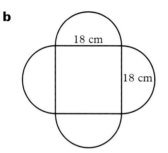

18 cm

18 cm

• d

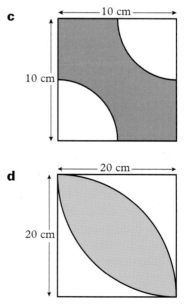

3 cm   3 cm

3 cm   3 cm

2 Find the shaded areas.

a

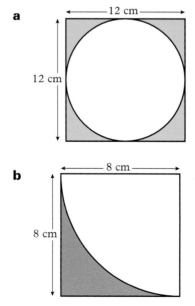

12 cm

12 cm

c

10 cm

10 cm

b

8 cm

8 cm

d

20 cm

20 cm

3 The goal area on an ice hockey rink is a semi-circle.
The diameter is 3.6 m.
Find the area.

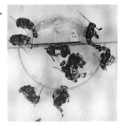

4 This table is a rectangle with a
semi-circle at each end.
Find the total area when all of
the table is being used.

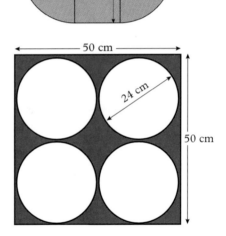

60 cm

140 cm

5 A machine cuts circular discs from
sheets of metal.
The sheets are 50 cm by 50 cm square.
Each circle has a diameter of 24 cm.
  a What is the area of the wasted
    metal left from each sheet after the
    4 discs have been cut out?
    In 1 day, 3000 sheets pass through
    the machine.
  b How many discs could be made
    from the wasted metal?

50 cm

24 cm

50 cm

6 Anna has drawn these three circles.

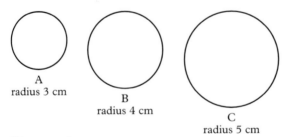

A
radius 3 cm

B
radius 4 cm

C
radius 5 cm

She sees that
      radius of A + radius of C = 2 × radius of B
Anna thinks that this means that
      area of A + area of C = 2 × area of B
Is Anna correct? Show all your working.

Exercise 3:6

Give all answers in this exercise correct to 1 dp.

1 This display board is a rectangle with a
semi-circle on the top.
  a Find the display area.
     There is beading all around the edge of
     the board.
  b Find the length of the beading.

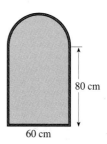

80 cm

60 cm

2 Paul has hit a cricket ball through this circular
window.
The diameter of the window is 56 cm.
The glass to repair the window costs 50 p for
250 cm².
The beading around the window also needs to
be replaced.
This costs £1.50 per metre.
How much does Paul have to pay for the
materials to fix the window?

3 A farmer has 240 m of fencing to make a pen for his chickens.
What area will he be able to enclose
  a if the pen is square,
  b if the pen is circular?

4 Find:
  a the radius of these circles   **b** the circumference of these circles.
  Hint: you need to change the formula to get the radius from the area.

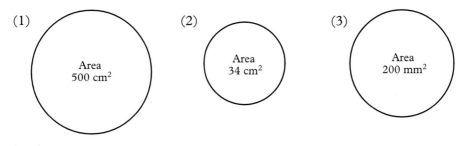

(1) Area 500 cm²

(2) Area 34 cm²

(3) Area 200 mm²

Give your answers correct to 1 dp.

1 Find the perimeter of a circle with
  a diameter 36 cm
  b radius 29 cm

2 Find the area of a circle with
  a radius 32 cm
  b diameter 24 cm

3 Find the diameter of a circle with circumference 340 cm.

4 The wheel of a bike has a diameter of 50 cm.
  a Find the circumference of the wheel in metres.
  b How far has the bike moved when the wheel has made 1 turn?
  c How many complete turns will the wheel have made when the bike
    has travelled 1 km?

5 This rectangular patio has a circular
  pond at one end.
  The diameter of the pond is 150 cm.
  Find the area of the paved part of
  the patio.

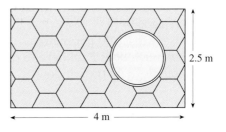

2.5 m

4 m

6 Jane is making lavender bags for the school fayre.
  She uses a circle of linen of diameter 20 cm for each bag.
  a What is the area of each circular piece of linen?
  Jane starts with a rectangular piece of linen 240 cm by 98 cm.
  b What is the maximum number of circles that she can cut out?
  c What percentage of linen will she waste?

7 Find the area of this shape.

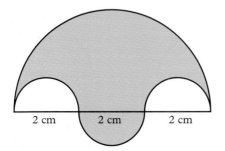

2 cm   2 cm   2 cm

Give your answers to 1 dp unless the question tells you to do something else.

1 Sonya has made a trundle wheel. She used a diameter of 30 cm.
A trundle wheel is meant to have a circumference of 1 m.
   a How many times should the trundle wheel rotate to cover 1 km?
   b How many times will Sonya's trundle wheel rotate in 1 km?
Sonya uses her trundle wheel to measure 500 m. She counts 500 rotations.
   c How far has Sonya actually measured? Give your answer in metres to 2 dp.
   d What is the percentage error that Sonya has made?

2 The minute hand of this clock is 9 cm long.
   a How far does the tip of the minute hand travel in 1 hour?
The hour hand travels 3.1 cm in 1 hour.
   b How long is the hour hand?

3 The diameter of Gavin's car wheel is 56 cm.
   a How far has the car travelled when the wheel has turned through 1 revolution?
On the motorway Gavin is travelling at 110 km per hour.
   b How many revolutions does the wheel make in 1 minute?

4 This archery target is marked so that the diameters of the circles are in the ratio
       1 : 2 : 3
Find the ratio of the areas.

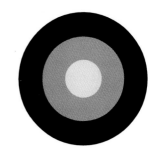

5 Charlotte is making a telephone table.
This is her design sketch.
Find the area of wood that she needs.

15 cm   15 cm

70 cm

6 Tony has two cake tins.
These circles are the bases of the tins.
a Find the area of the smaller circle.
The larger circle has twice the area of
the smaller circle.
b Find the radius of the larger circle.

10 cm

7 Sally is building a circular fish pond.
She wants to have at least 6 fish.
Each fish needs 900 cm² of surface area.
a What is the least area that Sally's
pond can have?
b What is the radius of the pond that
has this least area?
Sally actually uses a radius of 1.1 m.
c How many fish can Sally have?

8 Tom is making circular place mats.
He starts with square pieces of cork of side 35 cm.
a What is the area of the largest place mat that Tom can make?
Tom wants to make a place mat with half the area of the largest one.
b What radius should he use for this smaller place mat?

9 Sajid wants to make a poster.
He wants it to be a semi-circle.
He needs the area to be 400 cm².
What diameter should he use?

- **Circumference** The **circumference** of a circle is the distance around the circle.
  The circumference depends on the distance across the circle.

  Diameter The distance across a circle is called the **diameter**.

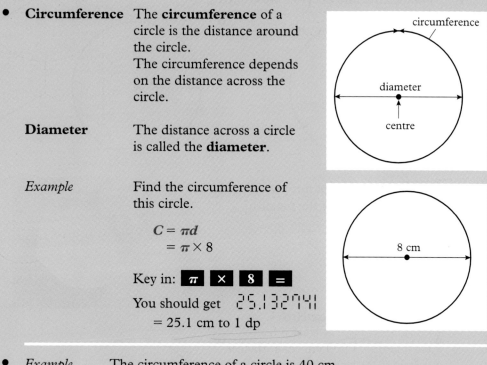

Example Find the circumference of this circle.

$$C = \pi d$$
$$= \pi \times 8$$

Key in: $\boxed{\pi}$ $\boxed{\times}$ $\boxed{8}$ $\boxed{=}$

You should get $25.1327\char"5C1$

$= 25.1$ cm to 1 dp

- *Example* The circumference of a circle is 40 cm. Find the diameter.

$$d = \frac{C}{\pi} \qquad \therefore \quad d = \frac{40}{\pi}$$

Key in: $\boxed{4}$ $\boxed{0}$ $\boxed{\div}$ $\boxed{\pi}$ $\boxed{=}$ 12.732395

$= 12.7$ cm to 1 dp

- **Radius** The **radius** of a circle is the distance from the centre to the circumference.
  The radius is half the diameter.

- **Area** The **area of a circle** depends on the radius of the circle.

  Example Find the area of this circle.

$$A = \pi r^2$$
$$= \pi \times 4^2$$
$$= \pi \times 4 \times 4$$
$$= \pi \times 16$$
$$= 50.3 \text{ cm to 1 dp}$$

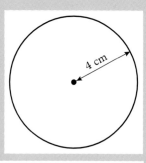

1 Find the circumferences of these circles.

a

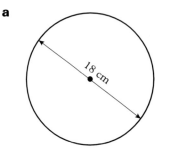

18 cm

b

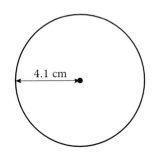

4.1 cm

2 The circumference of this pie dish is 82 cm.
What is the radius of the pie dish?

3 Find the areas of these circles.

a

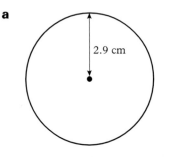

2.9 cm

b

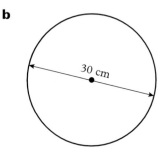

30 cm

4 Find the areas of these shapes.

a

7 cm

b

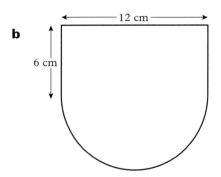

12 cm

6 cm

5 A circular flower bed has a diameter of 350 cm.
There is a path of width 80 cm around the flower bed.
Find the area of the path.

4 Statistics

Michelle Smith won three gold medals for Ireland in swimming events at the 1996 Olympics in Atlanta:

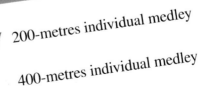

200-metres individual medley

400-metres individual medley

400-metres freestyle

1 Looking at data

This photo shows the men's 100 m final from the 1996 Olympic Games.
In 1956, the winning time in this event was 10.5 s.
The winning time in 1996 was 9.84 s.
This is an improvement of only 0.66 seconds, nearly $\frac{2}{3}$ of a second, in 40 years.

You need worksheet 4 : 1, **Olympic men's 100 m winners**.

In this section you are going to look at data from the 1996 Olympic Games.

You will collect information and draw diagrams to show the data that you have collected.

Statistical data is either discrete or continuous. You have used both types before.

Discrete data	When data can only take certain individual values it is called **discrete**.
Example	Shoe size is an example of discrete data. The values can only be 1, $1\frac{1}{2}$, 2, $2\frac{1}{2}$, etc. There are no shoe sizes between these.
Continuous data	Data is **continuous** when it can take *any* value in a certain range.
Example	The lengths of earthworms, the heights of pupils in Year 9, and the weights of hamsters are all examples of continuous data.

You are only going to look at continuous data in this section.

Exercise 4:1

W   You need worksheet 4 : 2, **Swimming: men's and women's 400 m freestyle**.
It shows the times of the competitors in the 1996 Olympic 400 m
freestyle competitions for men and women.

1   Look at the times in the men's preliminary rounds.
a   Copy this tally-table. Fill it in.

Time	Tally	Number of competitors
3 m 40 s but less than 3 m 50 s 3 m 50 s but less than 4 m 00 s 4 m 00 s but less than 4 m 10 s 4 m 10 s but less than 4 m 20 s 4 m 20 s but less than 4 m 30 s 4 m 30 s but less than 4 m 40 s 4 m 40 s but less than 4 m 50 s		

b   Draw a bar-chart to show this information.

2   Look at the times in the women's preliminary rounds.
a   Draw a tally-table for this information.
         Use the same groups as for the men.
b   Draw a bar-chart to show this information.

3   Look at your two bar-charts.
         Describe the differences between the two bar-charts.

4   Now look at the times in the A and B finals for the men and the women.
a   Draw new tally-tables for both the men's and the women's events.
         Include all the preliminary rounds and the finals.
b   Draw new bar-charts to show the information.
c   What do you notice about the new bar-charts?

● **5**   Draw better bar-charts to show the information on the worksheet.
         What do you need to do to draw better charts?
         What makes them better?
         Compare the bar-charts that you draw with each other and describe
         the differences between them.

2 Averages and range

A famous politician once said, 'We want everyone to be better than average'.

Why is this impossible?

There are 3 different types of average.

Mean	To find the **mean** of a set of data: (1) Find the total of all the data values. (2) Divide by the number of data values.
Mode	The **mode** is the most common or most popular data value. This is sometimes called the **modal value**. There can be more than one mode.
Median	To find the **median** put all the data values in order of size. The **median** is then the middle value. If there is an even number of data values, the median is the mean of the two middle values.

Exercise 4:2

1 These are the weights and heights of 5 sprinters.

Weight (kg)	85	91	74	68	82
Height (cm)	170	190	185	178	188

These are the weights and heights of 5 weightlifters.

Weight (kg)	104	108	112	107	104
Height (cm)	168	188	187	175	182

a Find the mean weight and height of the 5 sprinters.
b Find the mean weight and height of the 5 weightlifters.
c Do you think that the weightlifters would make good sprinters? Explain your answer.

2 These are the silver medal and bronze medal throws in the same seven men's Olympic javelin competitions.
They are rounded to the nearest metre.

Gold	90	90	95	91	89	84	90
Silver	89	90	88	90	86	84	87
Bronze	87	84	87	87	84	83	83

a What is the mode of the gold distances?
b What is the mode of the silver distances?
c What is the mode of the bronze distances?

The distances in the men's javelin in the 1996 Olympics were:

Gold: 88 m
Silver: 87 m
Bronze: 87 m

d If you include these distances, do they change any of the modes?

3 Here are the number of Olympic medals won by Italy from 1972–1996.

18   13   15   32   14   19   34

Find the median number of medals won by Italy.

4 These are the winning times in the women's 100 m for the past few Olympics.
They are given to the nearest tenth of a second.

11.4   11.0   11.1   11.1   11.1   11.0   10.5   10.8

Find the median time.

5 Write down the advantages and disadvantages of each of the three types of average.

Example

The mean height of 4 men is 185 cm.
Jim is 181 cm high. What is the mean height of the 5 men including Jim?

For the 4 men, the total of their 4 heights is
$$185 \times 4 = 740 \text{ cm}$$
For the 5 men, the total height is
$$740 + 181 = 921 \text{ cm}$$
Mean height of the 5 men $= 921 \div 5 = 184.2$ cm

6 The mean weight of 8 people is 60 kg. Alex weighs 52 kg.
What is the mean weight of the 9 people including Alex?

7 There are 29 pupils in 9X. Jason is away when there is a test.
The mean test score for the other 28 pupils is 76.5
Jason takes the test late and gets 86.
What is the new mean score for 9X?

8 There are 33 pupils in set 1.
When 32 of them take a test, their mean score is 86.5
Rubina is absent for the test and takes it late.
The teacher tells her that the mean for the whole class is now $86\frac{2}{3}$.
How many marks did Rubina get in the test?

The range

Cars travel a total of 270 000 million kilometres in the UK every year. The average for the UK, France, Germany and Spain is 255 000 million km.

This average does not tell the whole story. There is a big difference between the countries.

Germany's total is the biggest at 376 000 million. Spain's total is the smallest, at only 70 000 million.

The difference of 306 000 million km is called the **range** of the data.

Range	The **range** of a set of data is the biggest value take away the smallest value.

Exercise 4:3

1 These are the average ages that people live to in different countries. This is called 'life expectancy'.

Country	Life expectancy	
	Men	Women
UK	72	78
Angola	43	46
Brazil	62	68
Kenya	57	61
Iceland	75	80
Turkey	63	66

a Work out the mean and the range for men.
b Work out the mean and the range for women.

2 Look at this data. It is about 10 pupils in Year 11.
The data was taken when they were in Year 7, Year 9 and Year 11.
The heights are to the nearest centimetre.

Pupil	Sex	Y7 height	Y7 shoe size	Y9 height	Y9 shoe size	Y11 height	Y11 shoe size
1	F	145	2	154	3	160	4
2	F	153	3	160	5	167	5
3	F	137	1	147	3	154	4
4	F	141	2	156	4	165	6
5	F	149	2	160	3	170	4
6	M	131	3	157	5	165	7
7	M	151	5	159	7	175	9
8	M	141	3	155	5	171	8
9	M	150	4	165	5	180	10
10	M	129	3	135	4	155	7

a Find the mean and the range of the boys' heights in Year 7.
b Find the mode and the range of the boys' shoe sizes in Year 7.

3 a Find the mean and the range of the boys' heights in Year 9.
 b Find the mode and the range of the boys' shoe sizes in Year 9.

4 Do the same calculations for the girls.

5 a Which sex had the larger modal shoe size?
 b Which sex had the larger mean height?
 c Which sex had the larger range of heights?
 d Compare your answers for Year 9 with Year 7.
 Write down what you notice.

6 Do some calculations to see how the averages and ranges have changed by Year 11. Explain what has happened.
Use the questions in this exercise to help you.

Mean, mode, median and range for grouped data

It is not easy to work out averages for grouped data. Here are the boys' results in a Year 9 maths test. The results have been grouped in tens.

Mark	31 to 40	41 to 50	51 to 60	61 to 70	71 to 80	81 to 90	91 to 100
Number of boys	5	14	28	35	24	16	8

Look at the first column.
You can see that these boys scored between 31 and 40 but you do not know *exactly* what each of them scored.

To work out an **estimate** for the mean, you have to assume that all 5 of them scored the mark in the middle of the group.

This middle value is $\dfrac{31 + 40}{2} = 35.5$

In the same way you have to assume that the 14 people in the second column all scored the middle of that group which is $\dfrac{41 + 50}{2} = 45.5$

Now you can work out all the mid-points.
You can do a new table which looks like this.

Mark (mid-point)	35.5	45.5	55.5	65.5	75.5	85.5	95.5
Number of boys	5	14	28	35	24	16	8

Now you can work out the mean as if these were the scores that everybody got.

$$\text{Mean} = \frac{35.5 \times 5 + 45.5 \times 14 + 55.5 \times 28 + 65.5 \times 35 + 75.5 \times 24 + 85.5 \times 16 + 95.5 \times 8}{130}$$

$$= \frac{8605}{130}$$

$$= 66.2 \text{ (1 dp)}$$

This is only an **estimate** for the mean.

You have assumed that all the people in each group have scored the middle mark in each group.

This may not be true, so your mean may well be wrong!

When data is grouped you cannot tell which data value is the most common. You cannot find the mode. You can only say which group has the most values in it. This group is called the **modal group**. For the boys' test marks the modal group is 61 to 70 marks.

You cannot find the median either! You cannot write out the values in order so you can find the middle one. You can estimate the median value. This is done in the next section.

You can only estimate the range too!

An **estimate** for the range is the *biggest possible* value take away the *smallest possible* value.

For the boys' marks the biggest possible value is 100 and the smallest is 31.
An estimate for the range is 69 marks.

4

Exercise 4:4

1 Glen asked the pupils in his form how long it takes them to get to school.
Here are his results:

Time (mins)	1 to 5	6 to 10	11 to 15	16 to 20	21 to 25	26 to 30	31 to 35
Number of pupils	2	7	10	5	3	2	1

a Copy this table. Fill in the mid-points for each group of marks.

Mark (mid-point)	3						
Number of girls	2	7	10	5	3	2	1

b Work out an estimate for the mean. Give your answer to 1 dp.
c Write down the modal group for the times.
d Write down an estimate for the range of the times.

2 Eve asked all the people in her class how much money they spend
each month. Here are her results:

Amount (£)	£0–£9.99	£10–£14.99	£15–£19.99	£20–£24.99	£25–£50
Number of people	3	7	9	7	4

a Copy this table. Fill in the mid-points for each group.

Amount (£) (mid-point)					
Number of people	3	7	9	7	4

b Work out an estimate for the mean.
Give your answer to the nearest penny.
c Write down the modal group for the amount of money spent.
d Write down an estimate for the range of the amount of money spent.

3 Cumulative frequency

Have you ever wondered how exam grades are worked out?

How many marks does it take to pass or to gain a grade A?

Often this is worked out using percentages and cumulative frequency.
Examiners need to answer questions like,
'How many pupils scored less than 40%?' or,
'How many pupils scored more than 75%?'

To answer questions like these, a cumulative frequency table or a
cumulative frequency diagram is very useful.

Exercise 4:5

1 All the pupils in Year 9 took a maths test. Here are their marks as
percentages. They have been grouped into 10's.

Mark	Frequency	Mark	Frequency
1–10	3	51–60	24
11–20	6	61–70	14
21–30	11	71–80	6
31–40	13	81–90	3
41–50	18	91–100	2

a How many pupils took the test?
b How many pupils scored 20% or less?
c How many pupils scored 50% or less?
d How many pupils scored more than 70%?
e How many pupils scored more than 40%?

2   These are the marks for the Year 9 science test.

Mark	Frequency	Mark	Frequency
1–10	5	51–60	18
11–20	6	61–70	12
21–30	8	71–80	10
31–40	16	81–90	2
41–50	23	91–100	0

a   How many pupils scored 30% or less?
b   How many pupils scored 50% or less?
c   How many pupils scored more than 70%?
d   How many pupils scored more than 90%?
e   How could you *estimate* how many people scored 45% or less?

3   **a**   Copy this table.

Mark	Cumulative frequency	
10 or less	3	3
20 or less	9	3 + 6
30 or less	20	3 + 6 + 11
40 or less	33	3 + 6 + 11 + 13
50 or less	51	3 + 6 + 11 + 13 + 18
60 or less		
70 or less		
80 or less		
90 or less		
100 or less		

b   Finish the cumulative frequency column using the maths test data
    from question **1**.
    The final total should be 100 because 100 pupils took the test.
c   How many pupils scored 80 or less?
d   How many pupils scored 60 or less?
e   How many pupils scored *more* than 70?

**Cumulative
frequency**   **Cumulative frequency** is a running total.

Exercise 4:6

1 These are the marks in a science test.

Mark	Frequency	Mark	Cumulative frequency
1–10	5	10 or less	5
11–20	6	20 or less	11
21–30	8	30 or less	19
31–40	16	40 or less	35
41–50	23	50 or less	
51–60	18	60 or less	
61–70	12	70 or less	
71–80	10	80 or less	
81–90	2	90 or less	
91–100	0	100 or less	

a Copy this table.
b Finish the cumulative frequency column.
c How many pupils scored 60 or less?
d How many pupils scored *more* than 70?

2 This table shows the lifetimes in hours of a sample of 375 light bulbs.

Lifetime	Frequency	Lifetime	Cumulative frequency
201–400	56	400 or less	
401–600	124	600 or less	
601–800	101	800 or less	
801–1000	63	1000 or less	
1001–1200	31	1200 or less	

a Copy this table.
b Fill in the cumulative frequency column.
c How many bulbs lasted 600 hours or less?
d How many bulbs lasted 800 hours or less?
e How many bulbs lasted for more than 800 hours?

3 The table shows the results of a survey of 80 factory workers.
It shows their weekly take-home pay.

Wage	Frequency
121–140	11
141–160	17
161–180	24
181–200	19
201–220	9

a Draw a cumulative frequency table for this data.
b How many workers took home £180 or less per week?
c How many workers took home more than £160 per week?

Cumulative frequency curve	It is often useful to draw a curve from a cumulative frequency table.

The graph is called a **cumulative frequency curve**.
It allows you to estimate cumulative frequencies for points that are not at the ends of the groups.

A cumulative frequency curve is drawn with the values on the horizontal axis and the cumulative frequency on the vertical axis.
The points are always plotted at the **end** of each range.
The points are joined with a smooth curve.

Here is the light bulb data from Exercise 4 : 6 question **2**.

Lifetime	Cumulative frequency
400 or less	56
600 or less	180
800 or less	281
1000 or less	344
1200 or less	375

This is the cumulative frequency curve drawn from this data.

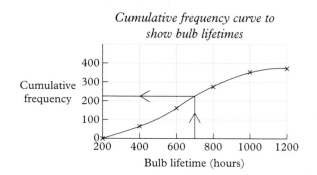

Cumulative frequency curve to show bulb lifetimes

Notice how the curve is joined back to the beginning of the first range in the original table. This is the point (200, 0).

Example

How many light bulbs lasted for less than 700 hours?

To estimate the number of bulbs that lasted less than 700 hours:
(1) Draw a line up from 700 on the horizontal axis to the curve.
(2) Draw across to the vertical axis.
(3) Read the value from this axis.
In this example this is approximately 225 bulbs.

Exercise 4:7

1 This is the maths test data you used in Exercise 4:5 question **1**.

Mark	Cumulative frequency	Mark	Cumulative frequency
10 or less	3	60 or less	73
20 or less	9	70 or less	89
30 or less	20	80 or less	95
40 or less	33	90 or less	98
50 or less	51	100 or less	100

a Copy these axes:

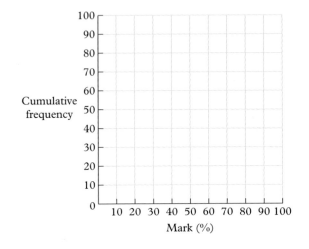

b Draw a cumulative frequency curve for this data.
c From your graph estimate the number of pupils scoring 45% or less.
d From your graph estimate the number of pupils scoring 73% or less.

2 You need to use the science test data from Exercise 4:6 question **1**.
a Use your cumulative frequency table to draw a cumulative frequency curve.
Use the same scale as you did for question **1**.
b From your graph estimate the number of pupils scoring 45% or less.
c From your graph estimate the number of pupils scoring 73% or less.
d Estimate how many pupils scored *more* than 45%.

3 These are the marks scored by a sample of 750 GCSE students. They are percentages.

Mark	Frequency	Mark	Frequency
1–10	20	51–60	147
11–20	48	61–70	84
21–30	65	71–80	64
31–40	124	81–90	24
41–50	157	91–100	17

a Draw a cumulative frequency table for this data.
b Draw a cumulative frequency curve for this data.

These are the marks needed for certain grades in the exam.

Grade	Mark
A	76%
B	70%
C	60%
D	52%

Estimate the following:
c How many pupils scored less than 52%?
d How many pupils scored less than 60%?
e How many pupils gained a grade D?
f How many pupils scored less than 70%?
g How many pupils gained a grade C?
h How many pupils gained a grade B?
i How many pupils gained a grade A?

It is possible to read other information from a cumulative frequency curve.

Median

The **median** is the middle data value.
To get an estimate of the median:
(1) Find the point halfway up the cumulative frequency axis.
(2) Draw a line *across* to the curve.
(3) Draw *down* to the horizontal axis.
(4) Read off the estimate of the median.

Lower quartile

The **lower quartile** is the value one quarter of the way through the data values.
To find the lower quartile:
(1) Find the point one quarter of the way up the cumulative frequency axis.
(2) Draw lines as you did for the median.

Upper quartile

The **upper quartile** is the value three quarters of the way through the data.

Interquartile range

The **interquartile range** is the difference between the upper quartile and the lower quartile. This tells you how spread out the central half of the data is.

4

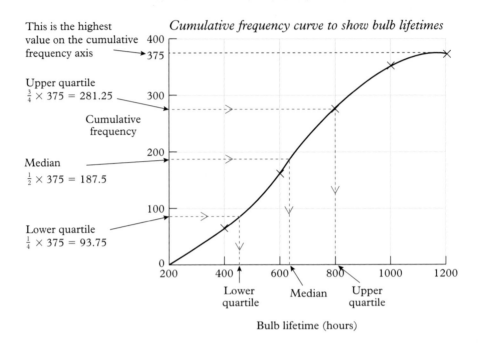

This is the highest value on the cumulative frequency axis ➝375

Upper quartile
$\frac{3}{4} \times 375 = 281.25$

Cumulative frequency

Median
$\frac{1}{2} \times 375 = 187.5$

Lower quartile
$\frac{1}{4} \times 375 = 93.75$

Cumulative frequency curve to show bulb lifetimes

Bulb lifetime (hours)

Median = 640   Lower quartile = 450   Upper quartile = 800
Interquartile range = 800 − 450 = 350

Exercise 4:8

1 This table shows the cost of 100 new cars.

Amount (£)	Frequency
5001–7000	15
7001–9000	34
9001–11 000	28
11 001–13 000	17
13 001–15 000	6

a Draw a cumulative frequency table for this data.
b Draw a cumulative frequency curve for this data.
c From your graph estimate the values of:
  (1) the median cost
  (2) the lower quartile cost
  (3) the upper quartile cost.
d Calculate the interquartile range in pounds.

2   Rajiv is checking the weights of two brands of crisps.
He checks 60 family bags of each brand. Here are his results:

Weight of crisps (g)	'Windsor' crisps number of packets	'Joggers' crisps number of packets
121–130	3	4
131–140	8	7
141–150	24	18
151–160	17	19
161–170	8	12

a   Draw a cumulative frequency table for both sets of data.
b   Draw a cumulative frequency curve for both sets of data.
    Draw them on the same graph.
c   Copy this table. Fill it in.

	'Windsor' crisps	'Joggers' crisps
Median		
Lower quartile		
Upper quartile		
Interquartile range		
Bags weighing < 145 g		
Bags weighing > 155 g		
Minimum weight of the heaviest 20 bags		

d   Write about the main differences between the two brands of crisps.

3   This table shows the amounts spent by 50 families on their summer holidays in 1990 and 1996.

1990 amount (£)	Frequency	1996 amount (£)	Frequency
101–250	7	101–250	10
251–400	18	251–400	14
401–550	16	401–550	12
551–700	7	551–700	8
701–850	2	701–850	6

a   Draw cumulative frequency graphs for both years.
b   For each year find (1) the median cost (2) the interquartile range
    in pounds.
c   Write about the differences between the two years.
    Use your graphs and calculations.

1 The graphs show the number of hours that men and women in the UK worked in 1990.

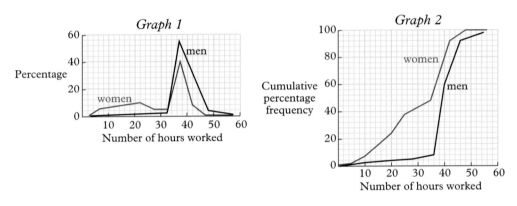

Look at *Graph 1*.
a Write down one thing that is similar about the pattern in the number of hours worked by men and women.
b Write down one thing that is different about the pattern in the number of hours worked by men and women.
Look at *Graph 2*.
c What percentage of women work for less than 35 hours per week?
d What percentage of men work for less than 35 hours per week?
e What percentage of women work for more than 45 hours per week?
f What percentage of men work for more than 45 hours per week?

2 Steve is doing a survey of how much people earn in a week.
He has drawn these bar-charts to show his results for 100 women and 100 men.

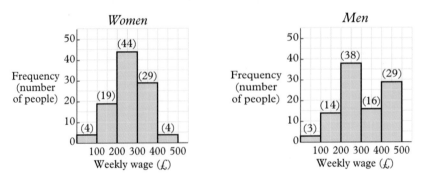

For parts **a** to **c** work out the answers for women and men.
a Work out an estimate for the mean weekly wage.
b Write down the modal group.
c Write down an estimate of the range of the weekly wages.
d Steve wants to tell people that women are paid as well as men. Can he use his data to back up this idea?
e Use the data to show Steve that he is wrong.

3 People who want to join the army take two tests.
The results of 60 people are shown on these cumulative frequency diagrams.

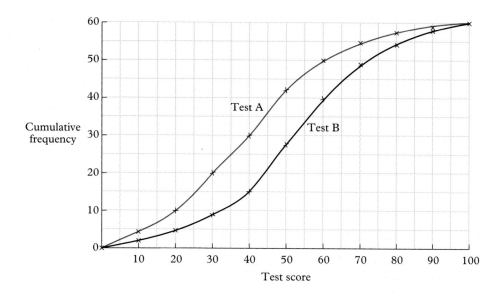

a Paul came 15th in Test A.
   Estimate his score.
b Ajay scored 60 in Test A.
   Out of the 60 people, in about what position was he?
c Liz scored 45 on Test B.
   Out of the 60 people, in about what position was she?
d The pass mark for Test A was 55%.
   How many people passed?
e The same number of people passed Test B.
   What was the pass mark for Test B?

4 These are the pulse rates of 400 company directors.

Pulse rates		Frequency
at least	below	
61	65	30
65	69	90
69	73	140
73	77	100
77	81	40

a Draw a bar-chart to show this data.
b Work out an estimate for the mean of these pulse rates.

1 200 pupils at Barking School took a maths test and a science test.
The results are shown on these cumulative percentage frequency curves.

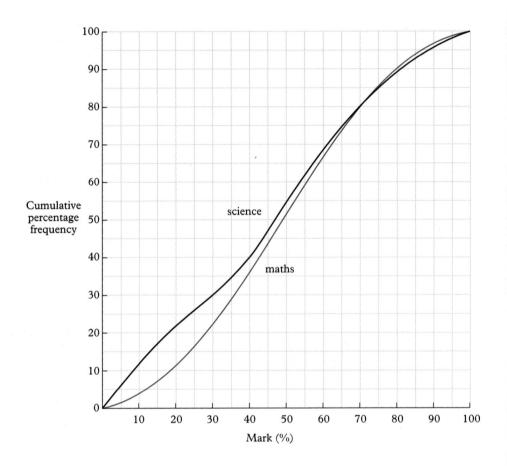

a Say whether these statements are true or false:
(1) Most pupils scored more than 40% in science.
(2) 35 pupils scored 40% or less in maths.
(3) Most people scored more than half marks in both tests.

b What is the median mark for maths?

c Is the median mark in science higher or lower than the median for maths?
Explain how you could tell this from the graph.

d Copy this sentence. Fill in the gaps.

.... % of pupils scored .... % or less in both tests.

- **Grouped data**

  It is not easy to work out averages for grouped data.
  To work out an estimate for the **mean** use the middle of the group and then work out the mean as if this was the value that everybody got.
  This is only an **estimate** for the mean.
  You cannot find the **mode**. You can only say which group has the most values in it.
  This group is called the **modal group**.
  An estimate for the **range** is the *biggest possible* value take away the *smallest possible* value.

- **Cumulative frequency**

  Cumulative frequency is a running total.

- **Cumulative frequency curve**

  It is often useful to draw a curve from a cumulative frequency table.
  The graph is called a **cumulative frequency curve**.
  It allows you to estimate cumulative frequencies for points that are not at the ends of the groups.
  A cumulative frequency curve is drawn with the values on the horizontal axis and the cumulative frequency on the vertical axis.
  The points are always plotted at the **end** of each range.
  The points are joint with a smooth curve.

- **Median**

  The **median** is the middle data value.
  To get an estimate of the median from a cumulative frequency curve:
  (1) Find the point halfway up the cumulative frequency axis.
  (2) Draw a line *across* to the curve.
  (3) Draw *down* to the horizontal axis.
  (4) Read off the estimate of the median.

- **Lower quartile**
  Upper quartile

  The **lower quartile** is the value one quarter of the way through the data values.
  The **upper quartile** is the value three quarters of the way through the data values.

- **Interquartile range**

  The **interquartile range** is the difference between the upper quartile and the lower quartile.

1 The data shows the amount of rain in millimetres that fell on each day in November.

3.5	16.4	6.4	3.7	14.2	8.9
22.9	2.9	7.8	18.9	0.1	2.6
9.4	14.2	4.5	11.6	15.9	6.1
13.7	13.9	3.1	2.5	5.6	1.4
6.9	4.1	17.9	19.2	10.7	7.2

a Draw a bar-chart to show this information.
You need to choose sensible groups for the data.
This graph shows the rainfall in April.

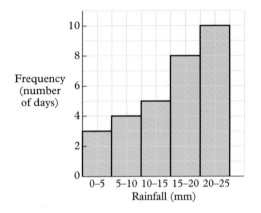

b Look at the bar-chart you have drawn and the one given in this question. Which one shows the wetter month? Explain how you can tell.
c Work out an estimate of the mean amount of rain in each month.

2 These are the pulse rates of 400 athletes.

Pulse rates		Frequency
at least	below	
55	57	70
57	59	110
59	61	120
61	63	90
63	65	10

a Draw a bar-chart to show this data.
b Work out an estimate for the mean of these pulse rates.
c Use the data in the table to draw a cumulative frequency table.
d Draw a cumulative frequency curve for the pulse rates.
e Write down the median pulse rate.
f Work out the interquartile range of the pulse rates.

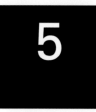

5 Accuracy

QUESTIONS

EXTENSION

SUMMARY

TEST YOURSELF

The number 10^{100} is called a googol. This is 1 with one hundred zeros.

$$10^{100} = \begin{aligned}&10\ 000\ 000\ 000\ 000\ 000\ 000\ 000\ 000\ 000\ 000\\ &000\ 000\ 000\ 000\ 000\ 000\ 000\ 000\ 000\ 000\ 000\\ &000\ 000\ 000\ 000\ 000\ 000\ 000\ 000\ 000\ 000\end{aligned}$$

The number 10^{googol} or $10^{10^{100}}$ is called a googolplex.

This is 1 with a googol zeros.

It takes Kerry $\frac{1}{4}$ second to write a zero and $\frac{1}{5}$ second to write the 1.

How many years would it take her to write a googolplex?

1 Rounding

Sarah has measured how long it takes for 60 swings of the pendulum.
She has worked out the time for one swing of the pendulum.
This is her answer:

Time for one swing = 1.036 251 seconds

This is not a sensible answer. She cannot measure time this accurately.
She needs to round off her answer.

◄◄REPLAY►

Examples   **1** Round the number 3.8 to the nearest whole number.

3   ↑
                         3.8   4

3.8 is nearer to 4 than to 3. It is rounded to 4.

2 Round the number 43 to the nearest ten.
43 is rounded to 40 to the nearest 10.

3 Round the number 8.39 to 1 dp.
8.39 is rounded to 8.4 to 1 dp.

Exercise 5:1

1 Round these numbers to the nearest whole number.
    a 14.85   **b** 30.6   **c** 79.5

2 Round these numbers to the nearest hundred.
    a 209   **b** 3829   **c** 999

3 Round these numbers to 1 dp.
    a 4.682   **c** 21.846   **e** 29.950
    b 10.147   **d** 47.981   **f** 50.97

4 Round these numbers to 2 dp.
    a 73.592   **c** 18.3397   **e** 9.895
    b 4.3554   **d** 0.3081   **f** 99.999

Sally is 153 cm tall correct to the nearest centimetre.

153 cm = 1.53 m
1.53 m is correct to the nearest centimetre.

Sally is 1.53 m tall correct to the nearest centimetre.

You can give a length in metres correct to 2 dp.
The length is then correct to the nearest centimetre.

Examples **1** Give these lengths correct to the nearest centimetre:
      a 347.9 cm   **b** 1.426 m

  a 347.9 cm correct to the nearest centimetre is 348 cm.
  b 1.426 m correct to the nearest centimetre is 1.43 m.

This rock weighs 1235 g correct to
the nearest gram.
1235 g = 1.235 kg
1.235 kg is correct to the nearest gram.

You can give a weight in kilograms correct to 3 dp.
The weight is then correct to the nearest gram.

  2 Give these weights correct to the nearest gram.
      a 41.2 g   **b** 0.8047 kg

  a 41.2 g correct to the nearest gram is 41 g.
  b 0.8047 kg correct to the nearest gram is 0.805 kg.

---

5 Give each of these lengths correct to the nearest centimetre.
    a 8.3 cm   **c** 4.132 m   **e** 0.167 m   **g** 13.205 m
    b 127.9 cm   **d** 11.037 m   **f** 12.325 m   **h** 5.009 m

6 Give these weights correct to the nearest gram.

a 56.8 g	**c** 7.2031 kg	**e** 3.5024 kg	**g** 0.0259 kg
b 129.4 g	**d** 1.2578 kg	**f** 0.3219 kg	**h** 2.1798 kg

Significant figure

In any number the first **significant figure** is the first digit which is not a 0
For most numbers this is the first digit.

The first significant figure is the red digit:
37.8   650   7.961   0.0538   0.002 004

Rounding to significant figures (sf)

To **round to any number of significant figures:**
a Look at the first unwanted digit.
b Use the normal rule of rounding.
c Be careful to keep the number about the right size.

A zero is only significant when it appears to the **right** of the first significant figure.

Examples

64.9 to 1 sf is 60   It is *not* 6!   0.0381 to 1 sf is 0.04
253.7 to 2 sf is 250   52 780 to 3 sf is 52 800
0.036 28 to 2 sf is 0.036   0.008 047 1 to 3 sf is 0.008 05

In the last example the 0 after the 8 is significant because it comes after the first significant figure.

7 Round these numbers to 1 sf.
a 48   **b** 264   **c** 37 821   **d** 0.070 94

8 Round these numbers to 2 sf.
a 657   **b** 12 089   **c** 499   **d** 98 800

9 Round these numbers to 3 sf.
a 9223   **b** 30 303   **c** 10 608   **d** 89 950

10 Round these numbers.
a 5.067 to 2 sf   **c** 20 785 to 3 sf   **e** 4555 to 2 sf
b 863.4 to 3 sf   **d** 0.059 82 to 2 sf   **f** 0.404 040 to 4 sf

11 How many significant figures are there in each of these numbers?
a 34.5   **c** 304   **e** 1100   **g** 0.08
b 17   **d** 270   **f** 0.319   **h** 0.901

It is easy to hit the wrong key when using a calculator.
It is a good idea to check that your answer is about right.

| Estimate | To get an **estimate** we round each number to 1 sf. |

Example

Work out 30.45 ÷ 8.7

Use a calculator to get 30.45 ÷ 8.7 = 3.5
Estimate: 30.45 ÷ 8.7 ≈ 30 ÷ 9

9 goes into 27 three times so 30 ÷ 9 ≈ 3

3 is near to 3.5 so the answer is probably right.

Exercise 5:2

In this exercise:
Work out the answers using a calculator.
Write down an estimate to check that each answer is about right.

1 **a** 21.28 ÷ 2.8   **c** 94.38 ÷ 3.9   **e** 129.6 ÷ 2.7   **g** 377.3 ÷ 4.9
   b 30.94 ÷ 9.1   **d** 43.07 ÷ 7.3   **f** 431.6 ÷ 8.3   **h** 495.6 ÷ 1.4

2 The table shows the number of people visiting a candle factory.

Week one	361
Week two	279
Week three	198
Week four	340
Week five	512

a What was the total number of visitors during the 5 week period?
b What was the mean number of visitors per week?

3 The table shows the prices and numbers of tickets sold for a concert.

How much money was taken?

Tickets sold	
Number	Price
218	£11.75
182	£9.25
46	£6.25

4 The number of pupils that went on a school trip was 1274.
Each coach carried 49 pupils. How many coaches were used?

Multiplying and dividing by numbers less than 1

Exercise 5:3

1 Rob and Ceri are doing this question.
$10 \div 0.5 = ?$
Rob says that the answer is 2 as numbers always get smaller when they are divided.
Ceri says that the answer is 20 as 0.5 is the same as a half and there are 20 halves in 10 whole ones.
a Which is the right answer?
b Can you see where the wrong answer came from?

2 Jen and Nathan are doing this question.
$12 \times 0.2 = ?$
Jen says the answer must be 24 because $12 \times 2 = 24$ and numbers always get bigger when they are multiplied.
Nathan says that the answer is 2.4 because 0.2 is less than 1 so the answer should be less than 12.
a Who is right?   **b** Why is the other person wrong?

In questions **3–6** decide on the correct answer without using a calculator.

3 $14 \times 0.1 = ?$
A 14   **B** 140   **C** 1.4   **D** 0.14

4 $15 \div 0.3 = ?$
A 5   **B** 0.5   **C** 50   **D** 1.5

5 $(0.5)^2 = ?$
A 25   **B** 2.5   **C** 0.10   **D** 0.25

6 $0.6 \div 0.3 = ?$
A 20   **B** 2   **C** 0.2   **D** 1.8

Check your answers using a calculator.

Game: Four in a line

This is a game for 2 players.
You need two colours of counters. Each player uses a different colour.

Player 1   Pick one whole number, × or ÷ and one fraction or decimal.
Example 2 × **0.3** Answer: 0.6
Work out your choice of question **without** your calculator.

Player 2   Do the same question **without** a calculator as a check.
If you get different answers check using a calculator.

Player 1   Cover up your answer on the board with a counter.
If you get an answer that is not on the board, you cannot
put a counter down.

Player 2   Now have a turn like player 1.

The first player to get four counters in a row in any direction is the winner.

1, 2, 3, 4, 5   × **or** ÷   0.1, 0.2, 0.3
6, 7, 8, 9, 10   0.4, $\frac{1}{2}$

100	3	2.1	0.7	90	14
18	70	0.5	40	1	60
0.9	0.2	2	12	8	10
1.2	0.8	30	5	0.6	1.5
1.8	4	0.1	50	6	0.3
0.4	16	1.6	20	1.4	80

Using a calculator

Kevin and two of his friends have bought a tent for £129.95
They are sharing the cost equally between them.
Kevin works out the three shares on his calculator like this: 129.95 ÷ 3
Here is the calculator display of the answer:

43.316667

Kevin says that he can get the calculator to round the answer to 2 dp.

Example

You can use the **fix mode** to round 43.316 667 to 2 dp.

Key in: **Mode** **7** **2**
        Fix mode   Number of decimal places

The display now shows 43.32   Answer: £43.32

To change the calculator back to normal key in **Mode** **9**

Exercise 5:4

For questions **1–2**, round your answers correct to 2 dp.
Set your calculator in fix mode by keying in **Mode** **7** **2**

1 Work these out:
  a £34.65 ÷ 4   **b** £21.99 ÷ 7   **c** £3.20 ÷ 6   **d** £123.50 ÷ 16

2 The school canteen sells 702 lunches. They take £658.45
What is the mean amount spent on one lunch?

For questions **3–4**, set your calculator in fix mode to round to 3 dp
by keying in **Mode** **7** **3**

3 12 identical balls weigh 1.1 kg. What does one ball weigh?
Give your answer correct to the nearest gram.

4 A pack of 8 chicken breasts weighs 1.75 kg.
What is the mean weight of one chicken breast?
Give your answer correct to the nearest gram.

Now put your calculator back to normal by keying in **Mode** **9**

2 Error in measurement

Peter is having some new furniture for his bedroom.
He is measuring the space to see if a new computer desk will fit.

Peter is using a measuring tape marked in centimetres.
He can only measure to the nearest centimetre.
Peter measures the space as 98 cm.

Lower and upper limits

The real length of the space for Peter's computer can be any value between 97.5 cm and 98.5 cm.
This can be shown on a number line.

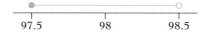

97.5 is the smallest length that rounds to 98.
97.5 is called the **lower limit**.
98.5 is halfway between 98 and 99.
The length cannot be exactly 98.5
but this is used as the **upper limit**.

Exercise 5:5

1 This catalogue has three different computer desks for sale.
The widths are correct to the nearest centimetre.

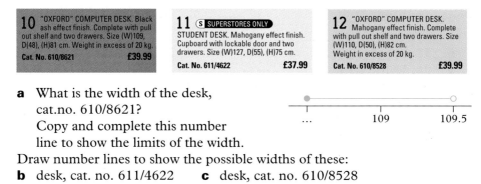

10 "OXFORD" COMPUTER DESK. Black ash effect finish. Complete with pull out shelf and two drawers. Size (W)109, D(48), (H)81 cm. Weight in excess of 20 kg.
Cat. No. 610/8621   £39.99

11 Ⓢ SUPERSTORES ONLY
STUDENT DESK. Mahogany effect finish. Cupboard with lockable door and two drawers. Size (W)127, D(55), (H)75 cm.
Cat. No. 611/4622   £37.99

12 "OXFORD" COMPUTER DESK. Mahogany effect finish. Complete with pull out shelf and two drawers. Size (W)110, D(50), (H)82 cm. Weight in excess of 20 kg.
Cat. No. 610/8528   £39.99

a What is the width of the desk, cat.no. 610/8621?
Copy and complete this number line to show the limits of the width.

Draw number lines to show the possible widths of these:
b desk, cat. no. 611/4622   **c** desk, cat. no. 610/8528

Example

Lisa has measured her pencil.
It is 8.3 cm correct to 1 dp.
What are the upper and lower limits of the length?

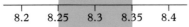

The lower limit is 8.25
The upper limit is 8.35

2 Lorna has measured her form room. It is 12.3 m long and 6.7 m wide correct to 1 dp.
   a What are the lower and upper limits of the length?
   b What are the lower and upper limits of the width?

3 The capacity of a fridge is 5.2 cubic feet correct to 1 dp.
  What are the upper and lower limits for the capacity of the fridge?

4 Write down the upper and lower limits for each of these:
   a The length of a pencil is 16 cm correct to the nearest centimetre.
   b The weight of a box is 23.6 lb correct to 1 dp.
   c The weight of a baby is 4.2 kg correct to one tenth of a kilogram.
   d A journey took 75 minutes correct to the nearest minute.

5 a The length of an envelope is 25 cm to the nearest centimetre.
     Show all the possible real lengths using a number line.
   b Mary has made a card. It is 24.5 cm long correct to 1 dp.
     Show all the possible real lengths using a number line.
   c Can Mary be sure that the card will fit in its envelope?
     Explain your answer.

6 The school record for the discus was measured as 33.51 m correct to the nearest centimetre.
  What are the upper and lower limits for the throw?

7 A scientist weighs a moon rock as 2.452 kg correct to the nearest gram.
  What are the upper and lower limits of the weight?

You can also have lower and upper limits for area and perimeter.

Example   The school playground is to be resurfaced and a fence is to be built.

The length and width of the playground have been measured correct to the nearest metre.

47 m

64 m

To find the **minimum** area you take the lowest values for the length and width.
For the **maximum** area you take the highest values.

Minimum area = 63.5 × 46.5   Maximum area = 64.5 × 47.5
                 = 2952.75 m²   = 3063.75 m²
The lower limit is 2952.75 m² and the upper limit is 3063.75 m²

Similarly, for the perimeter:
Minimum perimeter = 63.5 + 46.5 + 63.5 + 46.5 = 220 m
Maximum perimeter = 64.5 + 47.5 + 64.5 + 47.5 = 224 m
The lower limit is 220 m and the upper limit is 224 m.

Exercise 5:6

1 The sides of a rectangular field are measured correct to the nearest metre. The length is 329 m and the width is 251 m.
   a Write down the lower and upper limits of the length.
   b Write down the lower and upper limits of the width.
   c Find the minimum and maximum values of the area.
   d Find the minimum and maximum values of the perimeter.

2 The sides of a box are 6 cm, 8 cm and 10 cm correct to the nearest centimetre.
   Find the minimum and maximum values of the volume of the box.

3 Asha uses a metre rule to measure the length of her garden.
   She measures the length as 17 m exactly.
   The metre rule is only accurate to the nearest centimetre.
   Find the minimum and maximum values of the length of the garden.

4 The length of the side of a square
carpet tile is 30.0 cm correct to 1 dp.
Rudi lays 25 of the tiles in a row.
Find the lower and upper limits
for the length of the 25 tiles.

5 Greg is fitting kitchen units.
Each unit is 1.68 m long correct to the nearest centimetre.
He wants to fit 4 of these units into a space 6.73 m long.
Will the units fit? Explain your answer.

6 This rectangular sand pit has
dimensions 109 cm by 81 cm by
18 cm. All lengths are correct to
the nearest centimetre.
  a Find the lower and upper
  limits of the area of the
  sand pit.
  b Find the lower and upper
  limits of the volume of the
  sand pit.

7 The diameter of this paddling
pool is 183 cm correct to the
nearest centimetre.
  a Find the lower and upper
  limits of the area.
  b The depth of the pool is
  38 cm correct to the nearest
  centimetre.
  Find the lower and upper
  limits of the volume in litres.

• **8** If $p = 5.2$ cm, $r = 1.9$ cm and $s = 2.0$ cm correct to 1 dp, find the
largest possible value of:
  a $p + r + s$ **b** $p - r$ **c** $4r - p$ **d** $p \div r$

3 Standard form

This astronomer is looking at the nearest star to Earth.
Apart from the Sun, the star nearest to Earth is the very faint Proxima
Centauri which is 4.3 light-years away or 40 000 000 000 000 km.

Exercise 5:7

1 **a** Copy this number pattern and fill it in.

$$1^2 = 1 \times 1 \qquad = \ldots$$
$$10^2 = 10 \times 10 \qquad = \ldots$$
$$100^2 = 100 \times 100 \qquad = \ldots$$
$$1000^2 = 1000 \times 1000 = \ldots$$

   b Use the number pattern to write down the answer to $1\,000\,000^2$

2 **a** Copy this number pattern and fill it in.

$$4^2 = 4 \times 4 \qquad = \ldots$$
$$40^2 = 40 \times 40 \qquad = \ldots$$
$$400^2 = 400 \times 400 \qquad = \ldots$$
$$4000^2 = 4000 \times 4000 = \ldots$$

   b Use the number pattern to write down the answer to $4\,000\,000^2$

3 Write down the answers to these.

   a 3000^2 **c** 500^2 **e** $700\,000^2$

   b $20\,000^2$ **d** $60\,000^2$ **f** $8\,000\,000^2$

4 **a** Copy this table.
Fill in the (Number)^2 column using the patterns that you found in questions **1**, **2** and **3**.

Number	(Number)^2	Calculator display
1 000 000 2 000 000 3 000 000		

b Use a calculator to fill in the last column. Use the $\boxed{x^2}$ key.

For the first line, key in $\boxed{1}\ \boxed{0}\ \boxed{0}\ \boxed{0}\ \boxed{0}\ \boxed{0}\ \boxed{0}\ \boxed{x^2}$

Leave two more lines in your table for adding extra numbers.
c Look at columns two and three of your table.
Explain what you think the calculator display means.

5 **a** Add the numbers 4 000 000 and 5 000 000 to your table.
Fill in the table.
b What happens to the calculator display for these two numbers?

The calculator is using a method for writing numbers which are very large.

Standard form A very large number can be written as **a number between 1 and 10 multiplied by a power of 10**. This is called **standard form**.

$$4\,000\,000^2 \quad = 4\,000\,000 \times 4\,000\,000 = \mathbf{1.6 \times 10^{13}}$$

The calculator display shows 1.6^{13}

$\mathbf{1.6 \times 10^{13}}$ is standard form.
It means $1.6 \times 10\,000\,000\,000\,000 = 16\,000\,000\,000\,000$

16×10^{12} is not standard form because 16 is not between 1 and 10.

Example Write each calculator display as an ordinary number.

  a 8.1^{03} **b** 7.28^{05} **c** 2.06^{08}

  a $8.1 \times 10^3 = 8.1 \times 1000 = 8100$
  b $7.28 \times 10^5 = 7.28 \times 100\,000 = 728\,000$
  c $2.06 \times 10^8 = 2.06 \times 100\,000\,000 = 206\,000\,000$

Exercise 5:8

1 Write each calculator display as an ordinary number.

a 2.9^{04} **c** 9.0^{07} **e** 8.03^{01}

b 5.78^{06} **d** 1.57^{05} **f** 9.8374^{03}

2 Which of these numbers are not written in standard form?
Write the correct form where necessary.

a 3.05×10^2 **c** 0.6×10^5 **e** 9.2×10^3

b 49×10^6 **d** 495×10^9 **f** 3×10^6

3 These numbers are written in standard form.
Write them as ordinary numbers.

a 2.34×10^4 **c** 5.83×10^8 **e** 7.48×10^9

b 5.7×10^3 **d** 8.215×10^2 **f** 1.0×10^5

Examples

1 Write these numbers in standard form.

a 40 000 000 **b** 350 000 **c** 2 830 000

a $40\,000\,000 = 4$ multiplied by 10 seven times $= 4 \times 10^7$

b $350\,000 = 3.5$ multiplied by 10 five times $= 3.5 \times 10^5$

c $2\,830\,000 = 2.83$ multiplied by 10 six times $= 2.83 \times 10^6$

2 2.81^{03} on a calculator *must* be written as 2.81×10^3

Never use the calculator display method to write a number in standard form.

4 Write these numbers in standard form.

a 70 000 **c** 378 000 000 **e** 700 000 000

b 450 000 **d** 560 000 000 **f** 10 000 000 000

5 The average distance from the Sun to Earth is 93 000 000 miles.
Express this number in standard form.

6 This table shows the distances in kilometres of planets from the Sun.

Planet	Distance from the Sun (km)
Earth	1.5×10^8
Jupiter	7.78×10^8
Mars	2.28×10^8
Mercury	5.8×10^7
Pluto	5.92×10^9
Saturn	1.43×10^9
Uranus	2.87×10^9
Venus	1.08×10^8
Neptune	4.5×10^9

   a Write down each distance as an ordinary number.
   b Which planet is closest to the Sun?
     Explain how you can tell this from the standard form.
   c Which planet is furthest from the Sun?

7 Light travels about 9.5×10^{12} km in a year.
Write this as an ordinary number.

8 Light travels about 300 000 km in a second.
Write this in standard form.

9 There is a constant called Avogadro's number that is used in physics and chemistry. This is Avogadro's number:

      602 257 000 000 000 000 000 000

Write this number in standard form.

· ·

Standard form is also used for very small numbers.

Exercise 5:9

1 Copy this table.

$$10^4 = 10\,000$$
$$10^3 = 1000$$
$$10^2 = 100$$
$$10^1 = 10$$
$$10^0 = 1$$
$$10^{-1} = 0.1 \text{ or } \tfrac{1}{10} \text{ or } 1 \div 10^1$$
$$10^? = 0.01 \text{ or } \tfrac{1}{100} \text{ or } 1 \div 10^2$$
$$10^? = \ldots \text{ or } \tfrac{1}{1000} \text{ or } 1 \div 10^3$$
$$10^? = \ldots \text{ or } \ldots \text{ or}$$
$$10^? = \ldots \text{ or } \ldots \text{ or}$$

Use the patterns in the table to fill in the missing numbers.

Example   Write each calculator display as an ordinary number.

a 7.8^{-03} (display)   **b** 6.92^{-03} (display)   **c** 4.03^{-08} (display)

7.8^{-03} (display) means 7.8×10^{-3}

This means $7.8 \div 10^3 = 0.0078$ ($\div 10$ three times $0.\overset{3\,2\,1}{0078}$)

a $7.8 \times 10^{-3} = 7.8 \div 10^3 = 0.0078$
b $6.92 \times 10^{-5} = 6.92 \div 10^5 = 0.000\,069\,2$
c $4.03 \times 10^{-8} = 4.03 \div 10^8 = 0.000\,000\,040\,3$

---

2 Write each calculator display as an ordinary number.

a 5.8^{-03} **c** 4.72^{-02} **e** 4.53^{-08}

b 9.1^{-08} **d** 2.12^{-05} **f** 8.98^{-06}

3 Write each of these as an ordinary number.
a 2.6×10^{-3} **c** 5.68×10^{-2} **e** 5.39×10^{-8}
b 5.7×10^{-5} **d** 9.82×10^{-4} **f** 2.01×10^{-6}

Example   Write these numbers in standard form.
a 0.004 **b** $0.000\,354$ **c** $0.000\,000\,8$

a $0.004 = 4 \div 10^3 = 4 \times 10^{-3}$
b $0.000\,354 = 3.54 \div 10^4 = 3.54 \times 10^{-4}$
c $0.000\,000\,8 = 8 \div 10^7 = 8 \times 10^{-7}$

---

4 Write these numbers in standard form.
a 0.006 **c** 0.0007 **e** 0.0465
b 0.035 **d** $0.000\,56$ **f** 0.0039

5 Write these numbers in standard form.
a $0.000\,07$ **c** $0.000\,035$ **e** $0.000\,608\,2$
b $0.000\,000\,9$ **d** $0.000\,082$ **f** $0.000\,000\,000\,06$

6 This packet of paper contains
500 sheets.
Write down the thickness of
one sheet of paper.
Give your answer in standard form.

55 mm

7 The diameter of an atom is about 1.0×10^{-10}
Write this as an ordinary number.

Most calculators have an EXP key.
You can use this key to enter numbers in standard form into your
calculator.

Example Find the value of   **a** $3.4 \times 10^7 \times 6.9 \times 10^{-3}$
                                b $4.7 \times 10^9 \div 5.32 \times 10^{10}$

  a Keys to press:

| 3 | . | 4 | **EXP** | 7 | × | 6 | . | 9 | **EXP** | 3 | +/– | = |

Answer: 234 600 or 2.346×10^5 in standard form.

  b Keys to press:

| 4 | . | 7 | **EXP** | 9 | ÷ | 5 | . | 3 | 2 | **EXP** | 1 | 0 | = |

Answer: 0.088 345 8 or 8.8346×10^{-2} in standard form.

Exercise 5:10

Give all answers in standard form correct to 3 sf.

1 Work these out using the **EXP** key.
  a $2.4 \times 10^6 \times 8.1 \times 10^3$ **c** $6.19 \times 10^{-5} \div 4.3 \times 10^{-3}$
  b $7.64 \times 10^{-5} \times 2.3 \times 10^3$ **d** $7.09 \times 10^9 \div 3.2 \times 10^5$

2 This formula was discovered by Newton.
It gives the gravitational force (F) between two bodies.

$$F = \frac{GM}{R^2}$$

G is the gravitational constant and $G = 6.67 \times 10^{-20}$
M is the product of the masses of the two bodies.

Find the value of F when
a $M = 2.38 \times 10^{49}$ and $R^2 = 2.17 \times 10^{14}$
b $M = 3.56 \times 10^{51}$ and $R = 1.5 \times 10^8$

3 The average distance (R) of an object from the Sun is given by the formula

$$R = \frac{Gm}{V^2}$$

m is the mass of the sun $m = 1.993 \times 10^{30}$
G is the gravitational constant $G = 6.67 \times 10^{-20}$
V is the speed

Find the distance R when the speed V is equal to
a 25   **b** 16.4   **c** 3.4×10^2

4 A rocket travels a distance of 5.4×10^8 km in 9.2×10^3 hours.
Find the speed using the formula speed = distance ÷ time

5 Light travels at about 3×10^5 km/s.
Light takes about 8 minutes to reach the Earth from the Sun.
Find the distance between the Earth and the Sun in kilometres.

6 The frequency f and wavelength w of radio waves are connected by the
formula $fw = 3 \times 10^8$
a Find f when $w = 3.1$
b Find w when $f = 9.1 \times 10^7$

7 The mass of a hydrogen atom is about 1.67×10^{-24} g.
The mass of an oxygen atom is 2.7×10^{-23} g.
How many times heavier is the oxygen atom than the hydrogen atom?

1 Round these numbers:
   a 2.599 78 to 3 dp   **d** 24.3753 to 3 dp   **g** 3.999 to 2 dp
   b 3.576 to 1 dp   **e** 95 723 to 3 sf   **h** 0.040 067 1 to 4 sf
   c 3.582 to 2 sf   **f** 0.006 to 2 dp   **i** 0.005 82 to 2 sf

2 The price for printing a motif on
   a sweatshirt is:

   up to 5000 mm²   £2
   5000 mm² and over   £3

   This motif is 138 mm by 37 mm.

   a Estimate the area of the motif.
   b Use your estimate to find the cost of printing the motif.
   c Work out the actual area of the motif
   d Use your answer to **c** to find the actual cost of printing.
   e Explain why the two costs are different.

3 Peter and Mandy are doing this question: $(0.4)^2 = ?$
   Peter says that the answer is 1.6 because 0.4 is a decimal and 1.6 has a
   decimal as part of it.
   Mandy says that the answer is 0.16 because 0.4 is less than 1 so the
   answer must be less than 1.
   a Who is right?   **b** Why is the other person wrong?

4 The weights of four rowers are
   86.3 kg, 89.2 kg, 85.0 kg and
   93.9 kg correct to 1 dp.
   a Find the upper limit of their
      total weight.
   b Use your answer to **a** to find
      the maximum value of the
      mean weight.
   c Find the minimum value of
      the mean weight.

5 A room is 3.93 m long and 2.89 m wide correct to the nearest
   centimetre.
   Find the lower and upper limits of the area of the room.

6 A section of fencing is 2 m long correct to the nearest centimetre.
   What is the maximum length of 6 sections of fencing fixed together?

7 Pat does 30 mental calculations in 248 seconds to the nearest second.
What are the limits of the mean time for Pat to do one calculation?

8 A child's temperature rose from 37.7 °C to 38.8 °C correct to 1 dp.
Find the limits of the rise in temperature.

9 The rainfall for five separate months was 3.5 cm, 5.1 cm, 7.2 cm,
4.0 cm and 3.9 cm correct to the nearest millimetre.
  a Find the minimum total rainfall.
  b Find the minimum mean rainfall.
  c Find the maximum mean rainfall.

10 Rhian and Petra each have a pet stick insect.
They both say that the length of their stick insect is 5 cm to the nearest
centimetre. Does this mean that the insects are the same length?
Explain your answer.

11 This table shows the diameters of the planets in kilometres.

Planet	Diameter (km)
Earth	1.3×10^4
Jupiter	1.4×10^5
Mars	6.8×10^3
Mercury	4.9×10^3
Neptune	4.9×10^4
Pluto	2.4×10^3
Saturn	1.2×10^5
Uranus	5.2×10^4
Venus	1.2×10^4

  a Which is the largest planet?
  b Which is the smallest planet?
  c Which planet is between Mars and Pluto in size?
  d Which planet has a diameter about 10 times the diameter of Venus?
  e How many times bigger is Neptune than Mercury?
  f How many times bigger is Venus than Pluto?
  Give your answer in standard form.

12 A hydrogen atom has a mass of about 1.67×10^{-24} g.
What is the mass of 800 atoms?
Give your answer in standard form.

13 The radius of a circle is 5.35×10^9 cm.
Find the area of the circle. Give your answer in standard form correct to 3 sf.

1 The diameter of this bicycle wheel is 20 cm correct to the nearest centimetre.
Find the minimum number of revolutions needed to cover one kilometre.

2 The formula for the volume of a sphere is $V = \dfrac{4\pi r^3}{3}$

Find the lower and upper limits of the volume.
$r = 2.89$ cm correct to 2 dp.

3 A triangle has a base 3.91 cm and height 4.00 cm correct to 2 dp.
Find the lower and upper limits of the area.

4 If $x = 7.23$, $y = 6.08$ and $z = 3.00$ correct to 2 dp, find the smallest possible values of

  a $x + y$ **c** $y \div z$
  b $x - y$ **d** xyz

5 The mass of 1 atom of oxygen is 2.7×10^{-23} g.
What is the mass of 4×10^{28} atoms of oxygen?
Give your answer in standard form.

6 Light travels at a speed of 3×10^8 m/s.
Find how far light will travel in the following times.
Give your answer in standard form.

  a one hour   **b** one day   **c** one week
  d A light-year is the distance light can travel in one year.
     Find this distance giving your answer in standard form.
  e The distance of the nearest star from the Earth is 4.3 light-years.
     Find this distance in metres giving your answer in standard form.

- **Lower and upper limits** A space has length 98 cm correct to the nearest centimetre. The length of the space can be any value between 97.5 and 98.5 cm.
  This can be shown on a number line.

  |——————————————————————————————|
  97.5   98   98.5

  97.5 is called the **lower limit**.
  98.5 is called the **upper limit**.

- **Standard form** A very large number can be written as **a number between 1 and 10 multiplied by a power of 10**. This is called **standard form**.

  $4\,000\,000^2 = 4\,000\,000 \times 4\,000\,000 = \mathbf{1.6 \times 10^{13}}$

  The calculator display shows 1.6^{13}

  $\mathbf{1.6 \times 10^{13}}$ is standard form.
  It means $1.6 \times 10\,000\,000\,000\,000 = 16\,000\,000\,000\,000$

- *Example* Write each calculator display as an ordinary number.

  a 7.28^{05} **b** 7.8^{-03}

  a $7.28 \times 10^5 = 7.28 \times 100\,000 = 728\,000$
  b $7.8 \times 10^{-3} = 7.8 \div 10^3 = 7.8 \div 1000 = 0.0078$

- *Example* Write these numbers in standard form.
  a $40\,000\,000$ **b** $0.000\,354$

  a $40\,000\,000 = 4 \times 10^7$
  b $0.000\,354 = 3.54 \div 10^4 = 3.54 \times 10^{-4}$

- *Example* Find the value of   **a** $3.4 \times 10^7 \times 6.9 \times 10^{-3}$
  b $4.7 \times 10^9 \div 5.32 \times 10^{10}$

  a Keys to press:

  `3` `.` `4` `EXP` `7` `×` `6` `.` `9` `EXP` `3` `+/−` `=`
  Answer: 234 600 or 2.346×10^5 in standard form

  b Keys to press:

  `4` `.` `7` `EXP` `9` `÷` `5` `.` `3` `2` `EXP` `1` `0` `=`
  Answer: 0.088 345 8 or 8.8346×10^{-2} in standard form

1 Round these numbers:
   a 23.52 to nearest whole number   **b** 349 to nearest hundred
   c 35.572 to 1 dp   **e** 39.4729 to 3 dp   **g** 0.037 56 to 2 sf
   d 75.043 to 2 sf   **f** 27.462 to 3 sf   **h** 0.007 008 9 to 3 sf

2 **a** Give 5.649 m correct to the nearest centimetre.
    b Give 0.3782 kg correct to the nearest gram.

3 $20 \times 0.3 = ?$
  Pat says the answer must be 60 because $20 \times 3 = 60$ and numbers
  always get bigger when they are multiplied.
  Peter says that the answer is 6 because 0.3 is less than 1 so the answer
  should be less than 20.
    a Who is right?   **b** Why is the other person wrong?

4 Write each calculator display as an ordinary number.

   a 3.67^{03} **b** 6.293^{06} **c** 2.583^{-03}

5 These numbers are written in standard form.
  Write them as ordinary numbers.
   a 1.647×10^2 **b** 7.32×10^5 **c** 9.03×10^{-4}

6 John has measured the sides of his desk to the nearest centimetre.
  The length is 130 cm and the width is 52 cm.
   a Write down the lower and upper limits of the length and the width.
   b Find the minimum and maximum values of the perimeter.
   c Find the minimum and maximum values of the area.

7 **a** Estimate the area of this poster.
    Show the numbers you are using
    for your estimate.
   b Work out the actual area of the poster.

78.5 cm

44.5 cm

8 Sound travels 1.99×10^4 m each minute.
  The distance from the Sun to the Earth is about 1.5×10^{11} m.
  How long would sound take to travel from the Sun to the Earth if it
  were possible. (Sound cannot travel through a vacuum!)
  Use the formula   time = distance ÷ speed

6 Volume

QUESTIONS

EXTENSION

SUMMARY

TEST YOURSELF

Many scientists believe that global warming is melting the polar ice caps.

A study by the Inter-governmental Panel for Climate Change estimates that by the year 2070 global sea level could be as much as 71 cm higher than it is now.

Find out which parts of the UK would be underwater.

1   Volumes of prisms and cylinders

Gas is stored in holders like this.
The holder is a cylinder.
The height of the cylinder changes
as the volume of gas stored changes.

◄◄**REPLAY**►

Prism	A **prism** is a solid which is exactly the same shape all the way through. Wherever you cut a slice through the solid it is the same size and shape.
Cross section	The shape of the slice is called the **cross section** of the solid. A prism has a polygon as its cross section.
Cylinder	A **cylinder** is like a prism but it has a circle as its cross section.

Exercise 6:1

1 Look at these solids.
  a Write down the letters of the solids that have the same cross section all the way through.
  b Write down the name of the shape of the cross section for each solid in part **a**.

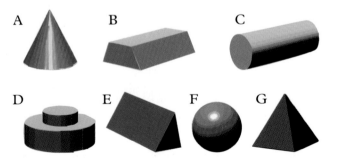

A   B   C

D   E   F   G

Volume of a prism	The volume of a prism or cylinder is **area of cross section × length**

Example

Find the volume of this prism.

Area of cross section = 25 cm²

Volume = 25 × 7
    = 175 cm³

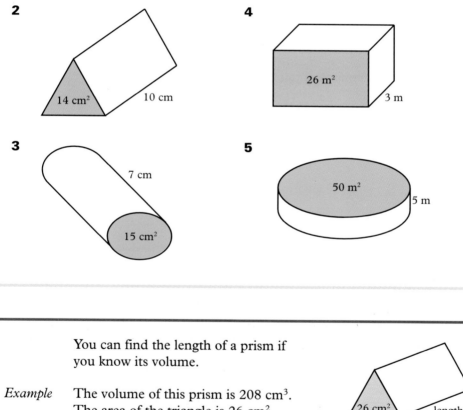

Work out the volumes of these solids.

2   14 cm²   10 cm

4   26 m²   3 m

3   7 cm   15 cm²

5   50 m²   5 m

You can find the length of a prism if you know its volume.

Example   The volume of this prism is 208 cm³.
The area of the triangle is 26 cm².

26 cm²   length

Volume = area of cross section × length

$\quad$ 208 = 26 × length $\qquad$ *The inverse of ×26 is ÷26*

208 ÷ 26 = length $\qquad$ *Divide each side by 26*

$\quad$ length = 8 cm

Find the lengths of these shapes.

6

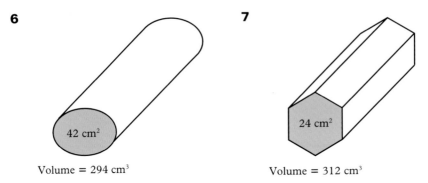

42 cm²

Volume = 294 cm³

7

24 cm²

Volume = 312 cm³

Sometimes you have to work out the area of the cross section first.

◄◄REPLAY►

Example   Find the volume of this prism.

Area of cross section = $\dfrac{\textbf{base} \times \textbf{height}}{\textbf{2}}$

$= \dfrac{6 \times 3}{2}$

$= 9 \text{ cm}^2$

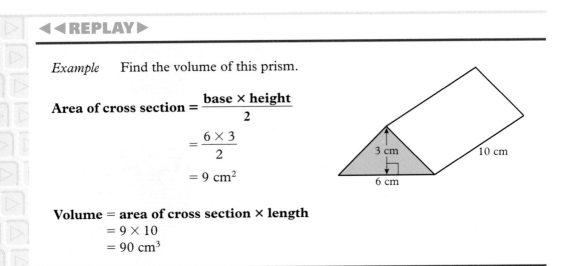

3 cm

10 cm

6 cm

Volume = area of cross section × length
= 9 × 10
= 90 cm³

Exercise 6:2

Work out the volume of each of these prisms.

1

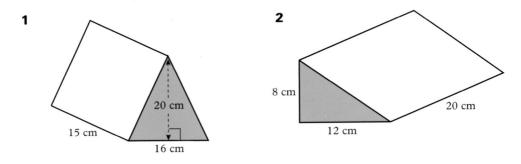

20 cm

15 cm

16 cm

2

8 cm

12 cm

20 cm

3

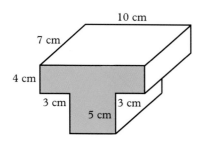

10 cm

7 cm

4 cm

3 cm 3 cm

5 cm

4

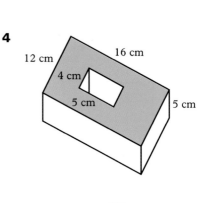

12 cm 16 cm

4 cm

5 cm 5 cm

5 Kiran is making a ramp for a wheelchair.
What volume of concrete does he need?

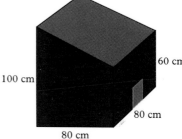

18 cm

110 cm 150 cm

6 Harry has built a coal bunker.
The picture shows the lengths of the sides.
What is the volume of Harry's coal bunker?

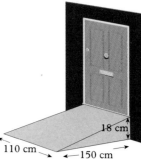

60 cm

100 cm

80 cm

80 cm

Find the lengths marked with a letter.

7

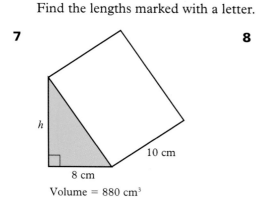

h

10 cm

8 cm

Volume = 880 cm³

8

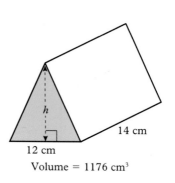

h

14 cm

12 cm

Volume = 1176 cm³

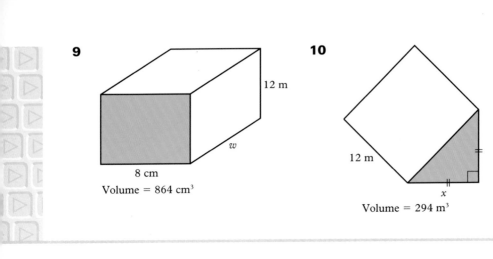

9 12 m / w / 8 cm

Volume = 864 cm³

10 12 m / x

Volume = 294 m³

Volume = area of cross section × length

The cross section of a cylinder is a circle.
The area of a circle is $\pi \times$ radius $\times$ radius

Example   **Area of cross section** $= \pi \times$ **radius** $\times$ **radius**
$$= 3.14 \times 5 \times 5$$
$$= 78.5 \text{ cm}^2$$
Volume of cylinder $= 78.5 \times 12$
$$= 942 \text{ cm}^3$$

10 cm

12 cm

Exercise 6:3

Work out the volumes of these cylinders.
Give your answers correct to 1 dp.

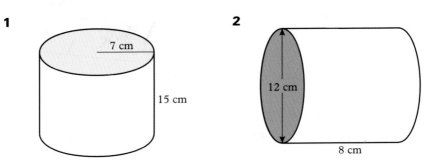

1   7 cm / 15 cm

2   12 cm / 8 cm

3

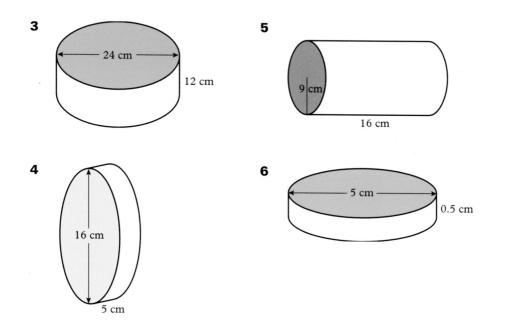

24 cm

12 cm

5

9 cm

16 cm

4

16 cm

5 cm

6

5 cm

0.5 cm

Exercise 6.4

1 Dale is buying corned beef.
These tins cost the same.
Dale has to decide which of
the two tins is the better value.
   a Find the volume of each tin.
   b Which tin gives the better value?

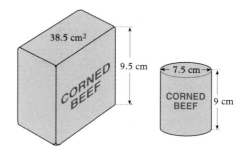

38.5 cm²

9.5 cm

CORNED
BEEF

7.5 cm

CORNED
BEEF

9 cm

2 Sinita has two cake tins. They are both 10 cm high.
One tin has a circular base. The other tin has a square base.

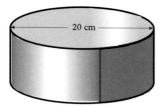

20 cm

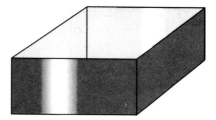

   a Find the volume of the tin with the circular base.
   b The two tins have the same volume.
   Find the length of the side of the square for the square based tin.
   Give your answer correct to 3 sf.

3 This cylindrical bottle contains water
to a depth of 15 cm.
The diameter of the bottle is 10 cm.

The water is poured into ice cube moulds.
The cubes have sides 3 cm long.

  a Find the volume of the water in the bottle.
  b How many moulds will the water fill?

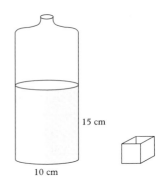

15 cm

10 cm

4 A wire has a circular cross section.
The diameter is 0.4 cm.
The wire is 150 cm long.
Find the volume of the wire correct to 1 dp.

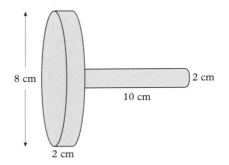

5 This peg is made from two
cylinders.
One cylinder has diameter
8 cm and height 2 cm.
The other cylinder has height
10 cm and diameter 2 cm.

Find the volume of the peg
correct to 1 dp.

8 cm

10 cm

2 cm

2 cm

6 These two cylinders have the
same volume.

  a Find the volume of cylinder A.
  b Find the length of cylinder B.
    Give your answers correct to 1 dp.

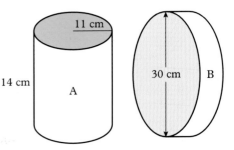

11 cm

14 cm

A

30 cm

B

7   A garden roller has a circular cross section.
The radius is 24 cm.
  a Find the area of the circle.
  b The volume of the roller is 226 080 cm³.
       Find the length of the roller correct to
       the nearest centimetre.

8   The photo shows the dimensions of
a cylindrical gas holder.
  a Find the number of litres of gas
       stored in the holder.
  b 1 million litres of gas are used
       during one day.
       How much does the height of the
       holder fall?
Give your answer correct to the nearest metre.

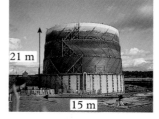

21 m

15 m

. . . .

Exercise 6:5

1   Each side of this cube is 1 cm.
  a What is the volume of the cube?
  b What is the length of a side in millimetres?
  c Use your answer to **b** to find the volume
       of the cube in mm³.
  d Copy and complete 1 cm³ = ... mm³

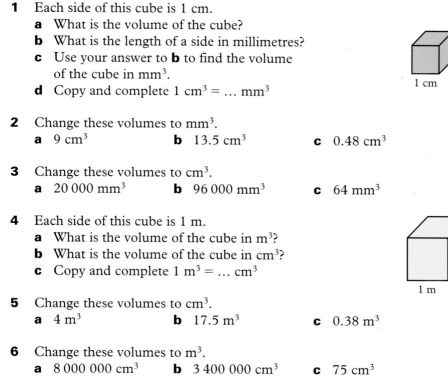

1 cm

1 cm

1 cm

2   Change these volumes to mm³.
  a 9 cm³   **b** 13.5 cm³   **c** 0.48 cm³

3   Change these volumes to cm³.
  a 20 000 mm³   **b** 96 000 mm³   **c** 64 mm³

4   Each side of this cube is 1 m.
  a What is the volume of the cube in m³?
  b What is the volume of the cube in cm³?
  c Copy and complete 1 m³ = ... cm³

1 m

1 m

1 m

5   Change these volumes to cm³.
  a 4 m³   **b** 17.5 m³   **c** 0.38 m³

6   Change these volumes to m³.
  a 8 000 000 cm³   **b** 3 400 000 cm³   **c** 75 cm³

It is easier to work in centimetres if you want
the volume in cm³ and in metres if you need
the volume in m³.
This saves you from having to convert the
units at the end.

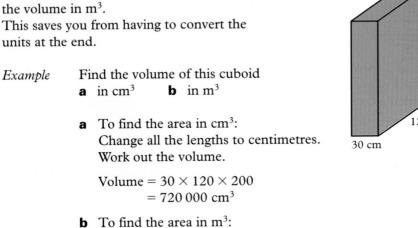

2 m

120 cm

30 cm

Example   Find the volume of this cuboid
a in cm³   **b** in m³

a To find the area in cm³:
Change all the lengths to centimetres.
Work out the volume.

Volume = 30 × 120 × 200
= 720 000 cm³

b To find the area in m³:
Change all the lengths to metres.
Work out the volume.

Volume = 0.3 × 1.2 × 2
= 0.72 m³

Find the volume of each of these solids
a in cm³   **b** in m³

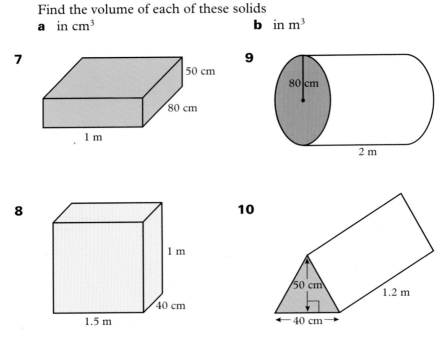

7

50 cm

80 cm

1 m

9

80 cm

2 m

8

1 m

40 cm

1.5 m

10

50 cm

1.2 m

40 cm

11 What is the capacity of this cylindrical water heater?
Give your answer in litres correct to the nearest litre.
Remember: 1 ml = 1 cm^3

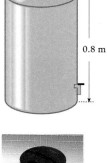

12 This drum has a capacity of 70 litres.
The height of the drum is 50 cm.
Find:
  a the capacity of the drum in cm^3
  b the area of the circular base in cm^2
  c the radius of the circular base
     in centimetres correct to 1 dp.

- -

Exercise 6:6 *Design a presentation box*

A company that makes cheese wants
a presentation box for three cheeses.

Each of the three cheeses is in the
shape of a cylinder.
The radius of each cheese is 2 cm
and the height is 3 cm.

The box must have these design features:

- It must not collapse when one or two cheeses are removed from the box.
- You must be able to see how many cheeses are in the box at any time.

Other things to consider are:

- The box needs to be simple to make.
- The box should not use too much card.
- Several of the boxes will need to be packaged together to send to the shops.
- The shops will want to stack the cheese boxes on shelves.

Design a suitable presentation box for the cheeses.
You will need to consider different arrangements for the cheeses.

Choose the final design and give reasons for your choice.

2 Another dimension

A line has one dimension.
This page has two dimensions.
Solid objects have three
dimensions (3-D for short).

The picture shows Dr Who's spaceship, the TARDIS.
This name stands for **T**ime **A**nd **R**elative **D**imension **I**n **S**pace.
The TARDIS travels through time –
the fourth dimension!
We only need to work in 1, 2 and 3 dimensions.

Paul is revising his work on area. He has written a list of the formulas
he needs to know.
This is his list:

$$\text{Area of rectangle} = \text{length} \times \text{width} \qquad = lw$$

$$\text{Area of triangle} = \frac{\text{base} \times \text{height}}{2} \qquad = \frac{bh}{2}$$

$$\text{Area of parallelogram} = \text{base} \times \text{height} \qquad = bh$$

$$\text{Area of circle} = \pi \times \text{radius} \times \text{radius} = \pi r^2$$

The red letters all represent lengths.
Area has two dimensions.
Any formula for area involves **length × length**
There may also be a number in the formula. The numbers are called
constants. In the formulas above, the 2 and π are constants.
A constant is not another dimension.

Example   Look at these formulas:

$$B = pr \quad C = 5qr \quad D = pqr \quad E = 7r^2$$

The letters p, q and r each represent a length in centimetres.
Which of these formulas could be for area?

$B = pr$ involves $p \times r$ which is **length** × **length**
$C = 5qr$ involves $5 \times q \times r$ which is constant × **length** × **length**
$D = pqr$ involves $p \times q \times r$ which is **length** × **length** × **length**
$E = 7r^2$ involves $7 \times r \times r$ which is constant × **length** × **length**

The formulas for B, C and E all have **length** × **length**
Each could be a formula for area.
The formula for D has too many lengths multiplied together.

Exercise 6:7

1 Look at these formulas:

$$P = bd \quad Q = b^2c \quad R = 7c^2 \quad S = 3bc$$

The letters b, c and d each represent a length in centimetres.
Which of these formulas could be for area?

2 In these formulas the letters r and t each represent a length in metres.

$$B = rt^2 \quad C = \frac{t^2}{4} \quad D = 2r \quad E = \frac{8rt}{5}$$

Which of these formulas could be for area?

In formulas a constant can sometimes be a letter.

Example   Look at these formulas: $A = 5w^2 \quad B = kw^2$

w represents a length and k is a constant.
Which formula could be for area?

$A = 5w^2$ involves constant × length × length
$B = kw^2$ also involves constant × length × length

Both formulas could be area formulas.

---

3 In these formulas the letters b, c and d each represent a length in centimetres, k is a constant.

$$S = 8bc \quad T = \frac{bc^2}{6} \quad W = kc^2 \quad X = \frac{cd}{k}$$

Which of these formulas could be for area?

4 The letters p, q and r each represent a length in centimetres, c and k are both constants.

$$A = kpq \qquad B = \frac{kp^2}{6} \qquad C = cq \qquad D = kpqr$$

Which of these formulas could be for area?

If there are two or more parts to a formula, each part must have **length** $\times$ **length** if the whole formula is for area.

Example

Look at these formulas.
The letters w, x and y each represent a length in centimetres.
k is a constant.

$$A = x^2 + 5wy \qquad\qquad C = 7x + w$$
$$B = wx - kxy \qquad\qquad D = ky^3 + 3wx^3$$

Which of these formulas could be for area?

In the formulas for A and B each part involves **length** $\times$ **length**.
Each could be a formula for area.
The formula for C has only one length in each part.
The formula for D has too many lengths multiplied together.

---

5 The letters d, e and f each represent a length in centimetres.
k is a constant.

$$S = kd^2 + 3ef \qquad\qquad W = 4d^2 + kf^2$$

$$T = def + 6d^3 \qquad\qquad X = \frac{de}{6} + kef$$

Which of these formulas could be for area?

6 The letters p, q and r each represent a length in centimetres.
c and k are both constants.

$$A = 4pq + r^2 \qquad\qquad C = \frac{4r^2}{3} + cpq$$

$$B = \frac{pq^2}{5} - kr^3 \qquad\qquad D = pq + kr$$

Which of these formulas could be for area?

7 Which of these formulas could be the formula for the area of the surface of a sphere of radius r?

$$A = \frac{3\pi r}{2} \qquad B = 3\pi r^3 \qquad C = 4\pi r^2 \qquad D = 5r^3$$

8 These wooden blocks are all the same shape but they are different sizes.

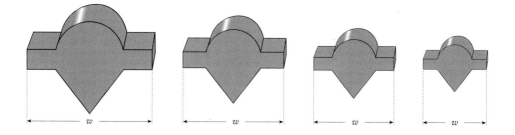

Look at these formulas:

$$A = kw^2 \qquad B = 4w \qquad C = \frac{w^3}{2}$$

w represents the width of a block, k is a constant.
Which formula could give the area of the face of one of the blocks?

9 This shape is called an ellipse:

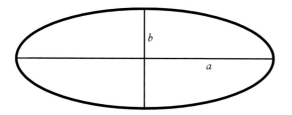

a Which of these formulas could be the formula for the area of the ellipse?

$$A = \pi a + \pi b \qquad B = \pi ab \qquad C = \pi a^2 b \qquad D = \pi ab + 4b$$

b Can you see a connection between the area of an ellipse and the area of a circle?

Diane has written a list of some of the formulas for volume:

Volume of a cuboid = length × width × height
Volume of a cylinder = π × radius × radius × height
Volume of a cube = length³

Volume has three dimensions.
Formulas for volume always have **length × length × length**

Example

Look at these formulas:

$$A = 7qr^2 \qquad B = 5r + kq \qquad C = 5pqr + kq^2r$$

The letters p, q and r each represent a length in centimetres.
k is a constant.
Which of these formulas could be for volume?

$7qr^2$ involves **length × length × length**
$5pqr$ and kq^2r both involve **length × length × length**
$5r$ and kq involve only **length**

Formulas A and C could be for volume.

Dimensions

Length has one **dimension**.
Formulas for length will only involve constants and lengths.
Area has two **dimensions**.
Formulas for area will only involve length × length.
They may include constants.
Volume has three **dimensions**.
Formulas for volume will only involve length × length × length.
They may include constants.

Exercise 6:8

1 A ball is placed inside a cube as shown.
Jacky has worked out a formula for the
space inside the cube.

Her formula is $V = \frac{3}{4}\pi r^2$

Use dimensions to explain why Jacky's
formula must be wrong.

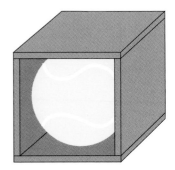

2 The letters p, q and r each represent a length in centimetres.
c and k are both constants.

$$A = 3pq + p^2 \qquad\qquad C = \frac{kp^2q}{3} - cpqr$$

$$B = \frac{2pq^2}{5} - kr^3 \qquad\qquad D = cpq + kr^2$$

Which of these formulas could be for volume?

Example   David has to decide which of these
formulas gives the perimeter of a shape.
He knows that perimeter is a length.
Length has just one dimension.

$P = 7x^2 + xy$
$Q = 3w + x$
$R = wx^2 + 5xy^2$

Each part of the formula for P is of the form **length × length**
Each part of the formula for R is of the form **length × length × length**
Only in the formula for Q is each part of the form **length**

David knows that $Q = 3w + x$ must be the formula that gives the
perimeter.

---

3 The letters f, g and h each represent a length in centimetres.
k is a constant.

$$A = 3f + 2h \qquad B = 3fg \qquad C = f^2 - kg^2 \qquad D = f + g + h$$

Which of these formulas could be for a perimeter?

4 Here are parts of the formulas you use for circles:

$$\pi d \qquad \pi r^2$$

d is the diameter and r is the radius of the circle.
a Which one do you use for circumference?
b Which one do you use for area?

5 These are formulas used for spheres.

$$\tfrac{4}{3}\pi r^3 \qquad 4\pi r^2$$

a Which one gives the surface area of a sphere?
b Which one gives the volume of a sphere?

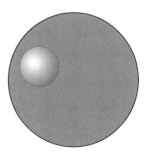

6 Which of these expressions could be
a formula for the area of this hexagon?
a $6a$ **b** a^3 **c** $2.6a^2$

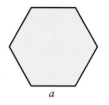

a

7 Which of these expressions could be a
formula for
a the area
b the perimeter of this running track?

$$wl + \frac{\pi w^2}{4} \qquad 2l^2 + \pi w \qquad 2l + \pi w$$

w

l

8 a, b, c and d are lengths on this torch.
Which of these expressions could be
a formula for the volume of the torch?
Explain your answer.

$$3bc + \pi d^2 \qquad bc^2 + \frac{3\pi d^3}{4} \qquad 5\pi d + 2a + c + b$$

9 Which of these formulas could be for:
a a length   **b** an area   **c** a volume?

$A = 8pq^2$

$D = pr - 7q^2$

$G = kp^2 - \pi r^2$

$B = kr^2 + \dfrac{pq}{4}$

$E = \dfrac{7p}{5} + r$

$H = \dfrac{kpq}{4}$

$C = 7r + \pi p$

$F = \dfrac{p^3 + r^3}{2}$

$I = \dfrac{kp}{3} + \dfrac{4q}{2} + \dfrac{\pi r}{5}$

p, q and r are lengths in centimetres, π and k are constants.

1 Ellen has made a device that collects rainfall.

The rain is collected in a rectangular tray 3 m long and 2 m wide.

It then runs into the measuring cylinder underneath.

The diameter of the cylinder is 40 cm.

What is the depth of water in the cylinder after 1 cm of rain has fallen into the tray?

Give your answer correct to the nearest centimetre.

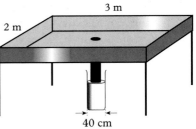

2 Richard is making glass fibre.

He starts with a cylinder of glass as shown in the diagram.

He spins this into glass fibre with a circular cross section of diameter 0.1 cm.

What length of glass fibre does he make?

Give your answer correct to the nearest metre.

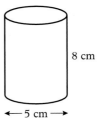

3 This diagram shows the cross section of a hosepipe.

Water flows through the pipe at a speed of 14 cm each second.

What volume of water will flow through the pipe in one minute?

Give your answer in litres correct to 1 dp.

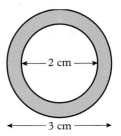

4 A firm that sells cat food wants to change the size of its tins.

The diagram shows the dimensions of the old tin.

a What is the volume of this tin correct to 1 dp?

b The new tin must have the same volume.

The new diameter is 8 cm.

What is the height of the new tin?

Give your answer correct to the nearest millimetre.

5 Which of these expressions could be for

a perimeter   **b** area   **c** volume?

r, l, w and h are lengths, π and k are constants

$\pi l + w$	$lw + 4h^2$	$2l - h$
$\pi l^2 h + 3rlw$	$5h^2 + wl$	$8w^3 - kl^2h$
$6\pi rl + klw$	$2\pi r + 3kl$	$hw^2 - \pi khl$

1   This engine has four cylinders.
Each cylinder has a diameter of 90 mm.
A piston moves the whole height of the cylinder.
The length of this piston stroke is 86 mm.
   a   Find the volume of one of the cylinders in cm³.
   b   Find the volume of the four cylinders in cm³.
          This value is called the capacity of the engine.
   c   Write the capacity of this engine in litres.
          Give your answers correct to one decimal place.

2   The diagram shows 25 cm of rubber tubing.
The inner radius is 2 cm and the
outer radius is 5 cm.
Find the volume of rubber used
to make the tubing.
Give your answer correct to 1 dp.

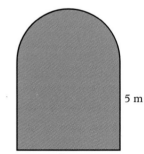

3   The picture shows the cross section of a tunnel.
The length of the tunnel is 356 m.
   a   What volume of earth was removed to
          make the tunnel?
          Give your answer correct to the nearest
          whole number.
   b   Lorries were used to transport the earth
          to a land fill site. Each lorry load was 8 m³.
          How many lorry loads of earth were transported?

5 m

5 m

4 d, h and l are lengths on a barn.
Which of these formulas could be the volume of the barn?
Explain your answer.

$$2d^2 + \frac{\pi hl}{4} \qquad 4d + 2\pi h + 5l \qquad d^2l + \frac{\pi h^2 l}{4}$$

5   **a**   Which of these formulas gives the
              surface area of this cube?

$$8b^3 \qquad 24b \qquad 24b^2$$

   b   What do the other two formulas give?
   c   Find the same formulas for this cube.

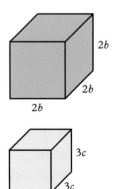

2b

2b

2b

3c

3c

3c

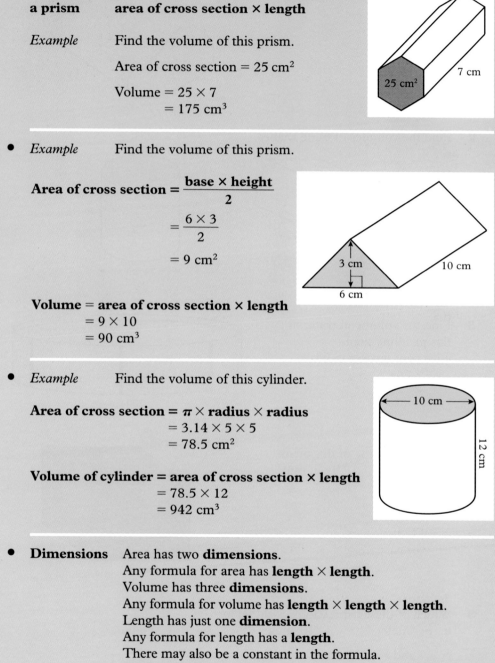

- **Volume of a prism**   The volume of a prism or a cylinder is
  area of cross section × length

  Example   Find the volume of this prism.

  Area of cross section = 25 cm²

  Volume = 25 × 7
  = 175 cm³

- *Example*   Find the volume of this prism.

  Area of cross section = $\dfrac{\textbf{base} \times \textbf{height}}{2}$

  $= \dfrac{6 \times 3}{2}$

  $= 9 \text{ cm}^2$

  Volume = area of cross section × length
  = 9 × 10
  = 90 cm³

- *Example*   Find the volume of this cylinder.

  Area of cross section = π × radius × radius
  = 3.14 × 5 × 5
  = 78.5 cm²

  Volume of cylinder = area of cross section × length
  = 78.5 × 12
  = 942 cm³

- **Dimensions**   Area has two **dimensions**.
  Any formula for area has **length × length**.
  Volume has three **dimensions**.
  Any formula for volume has **length × length × length**.
  Length has just one **dimension**.
  Any formula for length has a **length**.
  There may also be a constant in the formula.
  If there are two or more parts to a formula, then each part must
  have the same dimensions.

1   Find the volume of this solid.

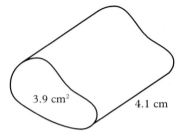

3.9 cm²   4.1 cm

2   Find the volumes of these shapes.

a

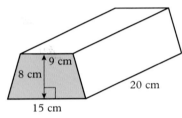

9 cm
8 cm
20 cm
15 cm

b

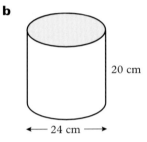

20 cm

24 cm

3   Find the volume of water in
this paddling pool.
Give your answer in litres.

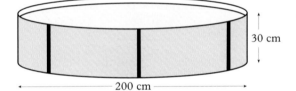

30 cm

200 cm

4   Find the volume of this prism
a   in cm³   **b**   in mm³

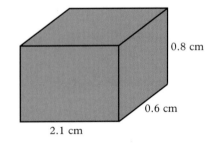

0.8 cm
0.6 cm
2.1 cm

5   Which of these expressions could be for
a   perimeter   **b**   area   **c**   volume?

p, q, r and s are lengths, π and k are constants.

$\pi q^2 + 4pr$ $pqr - kp^3$ $4p - kr + 2s$

$kpq + 3\pi pq$ $kp^2 + \pi s^2$ $7kpq - 5s^2$

7 Number revision

At sea level sound travels at 333.15 metres per second.

Light travels at 299 800 000 metres per second.

1000 m

The girl hears the bang about 3 seconds after she sees the flare explode.

1   ◄◄ **REPLAY** ►

71% of the surface of the earth is covered by water.

Example   This pie-chart has been divided
into 100 equal parts.
Each part is 1% of the pie-chart.

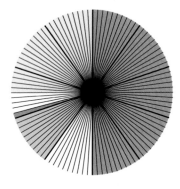

1% is coloured red.

$$1\% = \frac{1}{100} = 0.01 \text{ is coloured red.}$$

10% is coloured blue.

$$10\% = \frac{10}{100} = \frac{1}{10} = 0.1 \text{ is coloured blue.}$$

Exercise 7:1

1   **a**   How many parts of the pie-chart in the example are purple?
     b   What percentage is this?
     c   What decimal is this?
     d   What fraction is this?
     Give your fraction in its simplest form.

2   **a**   How many parts of the pie-chart in the example are yellow?
     b   What percentage is this?
     c   What decimal is this?
     d   What fraction is this?
     Give your fraction in its simplest form.

3 Cara has done 82% of her homework.
What percentage does she still have to do?

4 12% of year 9 at Stanthorne High go home to lunch.
What percentage stay at school for lunch?

5 The percentage of homes that have a telephone is 87%.
What percentage of homes do not have a telephone?

Here are two flow charts to help you to remember how to
work with percentages, fractions and decimals.

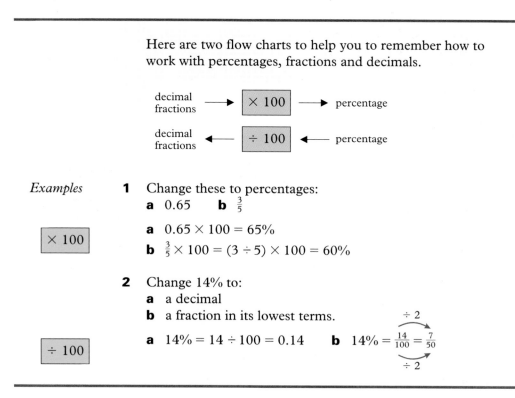

Examples

$\boxed{\times 100}$

$\boxed{\div 100}$

1 Change these to percentages:
  a 0.65   **b** $\frac{3}{5}$

  a $0.65 \times 100 = 65\%$
  b $\frac{3}{5} \times 100 = (3 \div 5) \times 100 = 60\%$

2 Change 14% to:
  a a decimal
  b a fraction in its lowest terms.

  a $14\% = 14 \div 100 = 0.14$ **b** $14\% = \frac{14}{100} = \frac{7}{50}$

Exercise 7:2

1 Change these decimal numbers to percentages.
  a 0.38   **b** 0.07   **c** 1.5   **d** 0.8   **e** 0.005

2 Change these fractions to percentages.
  a $\frac{3}{4}$ **b** $\frac{4}{5}$ **c** $\frac{11}{20}$ **d** $\frac{16}{25}$ **e** $\frac{1}{3}$

3 Change these percentages to decimals.
  a 79%   **b** 21%   **c** 7%   **d** 90%   **e** 30%

4 Change these percentages to fractions.
Give the fractions in their lowest terms.

a 70%	**d** 15%	**g** 26%	**j** $12\frac{1}{2}\%$	**m** $33\frac{1}{3}\%$
b 60%	**e** 55%	**h** 4%	**k** $37\frac{1}{2}\%$	**n** $66\frac{2}{3}\%$
c 5%	**f** 2%	**i** 24%	**l** $6\frac{1}{4}\%$	**o** $16\frac{2}{3}\%$

Robert, Chris and Fiona are
comparing their test marks.
Robert got $\frac{63}{75}$, Chris got $\frac{51}{60}$
and Fiona got $\frac{65}{80}$.
They are going to change
each mark to a percentage.
They can then see who got
the highest mark.

Example

Fiona got $\frac{65}{80}$. Change $\frac{65}{80}$ to a percentage.

To change a fraction to a percentage you multiply by 100.

$$\frac{65}{80} = \frac{65}{80} \times 100\% = (65 \div 80) \times 100\% = 81.25\%$$

Fiona got 81% correct to the nearest whole number.

Exercise 7:3

1 Change Robert's and Chris' marks to percentages.
Who got the higher mark?

2 Change these fractions to percentages.
Give your answers correct to the nearest whole number.

a $\frac{71}{80}$	**c** $\frac{4}{5}$	**e** $\frac{23}{36}$	**g** $\frac{27}{55}$	**i** $\frac{96}{120}$
b $\frac{16}{40}$	**d** $\frac{56}{60}$	**f** $\frac{15}{48}$	**h** $\frac{23}{28}$	**j** $\frac{112}{125}$

To work out what percentage one number is of another:
(1) Write the number as a fraction.
(2) Convert the fraction to a percentage.

Example A 35 g slice of wholemeal bread contains 3 g of fibre.

What percentage of the bread is fibre?

The fraction of bread that is fibre is $\dfrac{3}{35}$

Convert the fraction to a percentage.

$$\frac{3}{35} \times 100 = (3 \div 35) \times 100 = 8.5714 \ldots$$

$$= 8.6\% \text{ correct to 1 dp}$$

---

3 This is what Ellen did during one 24 hour period.

sleeping	8 hours	school	6 hours
eating	2 hours	watching TV	3 hours
at a disco	4 hours	other	1 hour

What percentage of her time did Ellen spend:
a sleeping   **c** at a disco
b watching TV   **d** eating?
Give each percentage correct to the nearest whole number.

4 Rajiv is looking at the populations of the villages in his local area.
This is his data.

Village	Barrow	Manley	Ashton	Sately
Population	928	735	1035	1256

a What is the total population?
b What percentage of the population live in Barrow?
c What percentage of the population live in Sately?
d What percentage of the population live in either Manley or Ashton?
Give each percentage correct to the nearest whole number.

5 George received £40 for his birthday.
He spent £24 and saved the rest.
a What percentage did George spend?
b What percentage did he save?

6 Theo works 40 hours each week.
His working hours are reduced to 38 hours a week.
What is the percentage reduction in the hours that Theo works?

2 Increase and decrease

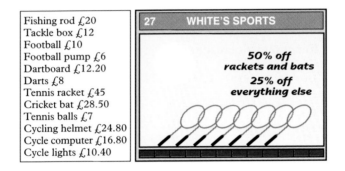

Fishing rod £20
Tackle box £12
Football £10
Football pump £6
Dartboard £12.20
Darts £8
Tennis racket £45
Cricket bat £28.50
Tennis balls £7
Cycling helmet £24.80
Cycle computer £16.80
Cycle lights £10.40

27   WHITE'S SPORTS

50% off
rackets and bats
25% off
everything else

Lisa wants to know how much she will save if she buys a dartboard.

Remember: 50% = $\frac{1}{2}$ 25% = $\frac{1}{4}$

Example

Lisa is buying a dartboard. It costs £12.20.
The shop has reduced the price by 25%.
a How much does Lisa save?
b How much does Lisa pay for the dartboard?

a Lisa knows that 25% = $\frac{1}{4}$ and that $\frac{1}{4}$ of £12.20 = £3.05
Lisa saves £3.05

b Lisa pays £12.50 − £3.05 = £9.45

Exercise 7:4

Use the prices given in the picture to answer the questions in this exercise.
For each item give:
a   the amount saved   **b**   the reduced price.

1   a tackle box   **6**   a football pump

2   a fishing rod   **7**   a cycling helmet

3   a tennis racket   **8**   a cricket bat

4   the tennis balls   **9**   a cycle computer

5   a football   **10** cycle lights

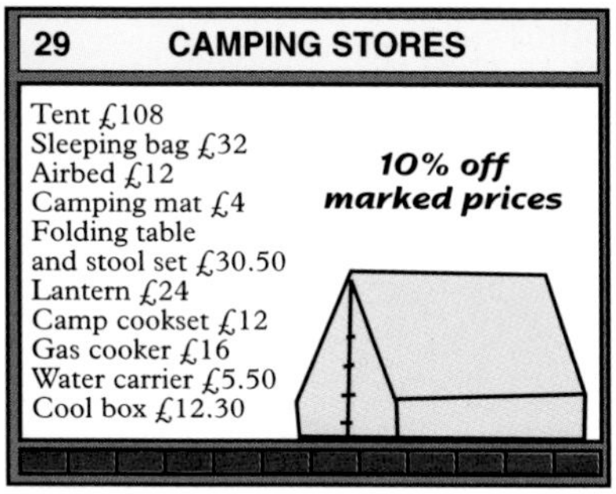

| 29 | CAMPING STORES |

Tent £108
Sleeping bag £32
Airbed £12
Camping mat £4
Folding table
and stool set £30.50
Lantern £24
Camp cookset £12
Gas cooker £16
Water carrier £5.50
Cool box £12.30

**10% off
marked prices**

Example

Alan is buying a tent. It costs £108.
The shop has reduced the price by 10%.
a How much does Alan save?
b How much does Alan pay for the tent?

a Alan knows that $10\% = \frac{1}{10}$ and that $\frac{1}{10}$ of $108 = \dfrac{108}{10} = 10.8$

Alan saves £10.80

b Alan pays £108 − £10.80 = £97.20

Exercise 7:5

Use the prices given in the picture for questions **1** to **8**.
For each item give:
a the amount saved   **b** the reduced price.

1 a sleeping bag   **5** an airbed

2 a gas cooker   **6** a cool box

3 a lantern   **7** a water carrier

4 a camp cookset   **8** a folding table and stool set

9 A mountain bike costs £120.
The shop is selling the bike at 10% off the full price.
What is the reduced price?

10 Huw gets £12 per week for his paper round.
The shopkeeper is going to increase his money by 10%.
How much will Huw get at this new rate of pay?

11 This box of tea contains 200 teabags.
A special offer pack contains 10% more.
   a How many extra tea bags is this?
   b How many tea bags in total are in
   the special offer pack?

12 A packet of sweets contains 200 g.
A special offer pack contains 20% more.
   a How many extra grams of sweets is this?
   b How many grams are there altogether in the special offer pack?

Sometimes you need a calculator to work out the percentage.

Example

Sian is buying a computer.
It costs £960.
The shop reduces the price by 12%.
How much does Sian save?

Sian needs to work out 12% of 960.

| Step 1 | Step 2 | Step 3 |

This can be written: $\dfrac{12}{100} \quad \times \quad 960$

Step 1   This changes the percentage to a decimal.
The decimal appears when you press ☒ at Step 2.

Step 2   'Of' is the same as multiply so use the ☒ key.

Step 3   You are finding the percentage of this amount.

Key in:

Step 1   Step 2   Step 3

[1] [2] [÷] [1] [0] [0] [×] [9] [6] [0] [=] 115.2

Answer: £115.20

Exercise 7:6

1 Find 12% of 300 g.

2 Find 18% of 250 cm.

3 Find 8% of 600 cm.

4 Find 45% of 120 g.

5 Find 34% of 350 m.

6 Find $12\frac{1}{2}$% of £150.

7 Find 15% of £450.

8 Find $9\frac{3}{4}$% of £420.

9 The volume of a standard size bottle of bath oil is 320 ml.
A special offer bottle contains 8% more.
   a How much more bath oil is in the special offer bottle?
   b What is the total amount of bath oil in the special offer bottle?

10 Helen's family went out for a meal.
The bill came to £32.50
They gave the waitress a $12\frac{1}{2}$% tip.
   a How much was the tip?
   b What was the total cost of the meal?

11 Alisha puts £125 in a savings account paying 5.9% interest per year.
The interest on the account is added at the end of each year.
Alisha does not intend to draw any money out.
Work out the amount she will have in the account after:
   a one year   **b** two years.

12 A shop had a sale. For each day of the sale the prices were reduced by
15% of the prices on the day before.
   a A sweater sold for £28 before the sale.
      Find its selling price after:
      (1) one day in the sale,
      (2) two days in the sale.
   b Find the selling price of each of these after two days in the sale.
      (1) A jacket with a price of £46.50 before the sale.
      (2) A pair of jeans with a price of £32.95 before the sale.

13 A car sales company estimates that a new car depreciates (loses value)
at 18% per year.
Calculate the value after two years of new cars selling at:
   a £8750   **b** £13 195

Example   Pierre wants a computer game for his birthday.
The computer game cost £54.
His parents agree to pay $\frac{3}{4}$ of the cost.
How much do Pierre's parents pay?

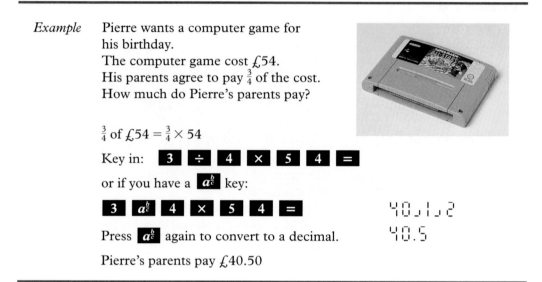

$\frac{3}{4}$ of £54 = $\frac{3}{4} \times 54$

Key in: | 3 | ÷ | 4 | × | 5 | 4 | = |

or if you have a $a\frac{b}{c}$ key:

| 3 | $a\frac{b}{c}$ | 4 | × | 5 | 4 | = |

Press $a\frac{b}{c}$ again to convert to a decimal.

$40\lrcorner1\lrcorner2$
40.5

Pierre's parents pay £40.50

Exercise 7:7

1 Use a calculator to find:

   a $\frac{3}{4}$ of £22   **c** $\frac{5}{8}$ of 400 people

   b $\frac{3}{5}$ of 375 houses   **d** $\frac{3}{20}$ of £6.80

2 Use a calculator to find these.
Give each answer correct to the nearest penny.

   a $\frac{2}{3}$ of £50   **c** $\frac{4}{11}$ of £5.60   **e** $\frac{5}{9}$ of £2.75

   b $\frac{3}{7}$ of £20   **d** $\frac{4}{5}$ of 32 p   **f** $\frac{7}{12}$ of 85 p

3 Danny is going to the sports centre. All prices are reduced by $\frac{1}{3}$.
Danny is going to the gym first and then he is going swimming.
Swimming usually costs £1.35 and the gym usually costs 75 p.
How much will Danny have to pay in total at the reduced price?

4 A flask can hold up to 750 ml of coffee. The flask is $\frac{7}{8}$ full.
How much coffee does it contain?

5 Barbara is buying a new tennis racket.
Shop A offers a 15% discount. Shop B offers to cut the price by $\frac{1}{8}$.
Which shop gives the lower price?
Show all your working.

Example   A coat is reduced by 15% in a sale.
Rachel buys the coat for £61.20.
Find the full price of the coat.

The *full* price is 100%.
The full price is reduced by 15%
to get the sale price.
The *sale* price is 100% − 15% = 85%

85% of the full price is £61.20

1% is $\dfrac{£61.20}{85}$ = £0.72

100% of the full price is £0.72 × 100 = £72

Exercise 7:8

1 A toy shop has reduced all its stock by 15% for its summer sale.
   a Find the old price of a model boat whose sale price is £6.80.
   b Find the old price of a toy garage whose sale price is £29.75.
   c Find the old price of a computer game whose sale price is £47.60.

2 The cost of a package holiday for a child is 70% of the adult price.
   The child's price is £336.
   a What is the cost for an adult?
   b What is the cost for two adults and three children?

3 The value of Paul's bike has decreased by 35% during the year.
   It is now valued at £96.20.
   What was the bike's value at the beginning of the year?

4 Jane has just received a phone bill for £59.86.
   This phone bill is 18% less than it was last time.
   How much was Jane's last phone bill?

5 Wasim has had his sports kit stolen. The insurance company only
   refunds 65% of the original value. Wasim receives £31.20
   How much did his sports kit originally cost?

Example Ruth put some money into a savings account.
The account pays 6% interest.
Ruth has £689 in her account after the interest has been paid.
How much did Ruth put into the account before the interest was paid?

Ruth's original savings are 100%.
The interest rate is 6%.
The original savings plus the interest
are 100% + 6% = 106%

106% is £689

1% is $\dfrac{£689}{106}$ = £6.50

100% = £6.50 × 100 = £650
Ruth put £650 into the account.

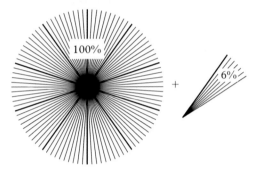

Exercise 7.9

1 Helen took a taxi home from the disco. She worked out that the fare plus a 10% tip would be £6.38. What was the cost of the fare?

2 Melissa bought some material and made some scarves. She sold the scarves for a total of £14.88. This was 20% more than she paid for the material. How much did she pay for the material?

3 Rita bought an old chair and repaired it. She then sold the chair for £36. This was 80% more than she originally paid for it. How much did Rita pay for the chair?

4 Kiran has had a pay rise of 4%. He now gets paid £8112 per year. What was he paid before the pay rise?

5 Sam opened a savings account one year ago with the money he received on his fourteenth birthday. The savings are now worth £61.48. Find the amount of money that Sam received for his fourteenth birthday. The rate of interest is 6% per year.

3 Ratio

There are the same number of girls as boys at Stanthorne High. The *ratio* of girls to boys is 1 : 1

◄◄ REPLAY ►

Here are six counters.

There are 2 red counters out of a *total* of 6 counters.

The *fraction* of the counters that are red is $\dfrac{2}{6} = \dfrac{1}{3}$

There are 2 red counters and 4 blue counters.

The *ratio* of red counters to blue counters is   2 : 4
You simplify ratios like fractions.   2 : 4 = 1 : 2

The ratio 1 : 2 tells you that there is 1 red counter for every 2 blue counters.
The ratio *compares* the number of red counters with the number of blue counters.

Ratio	**Ratio** is a measure of the relative size or quantity of things.

Exercise 7:10

Give your answers in their simplest form in this exercise.

1   Look at the red and blue counters in the example.
　　a   What fraction of the counters are blue?
　　b   What is the ratio of blue counters to red counters?

2 Liam is making this pattern with cubes.

    a Copy and complete these ratios for the colours in the pattern.
        (1) red : blue : yellow = 2 : ... : ...
        (2) blue : red : yellow = 1 : ... : ...
        (3) yellow : red : blue = ... : ... : 1
    b Liam continues his pattern. He uses 6 more reds.
        (1) How many more blues does Liam use?
        (2) How many more yellows does he use?
    c Liam uses 30 yellows altogether.
        (1) How many blues does he use?
        (2) How many reds does he use?

3 Sonja is making an orange drink.
She uses 4 parts water to 1 part
orange squash.
    a Write down the ratio of squash
        to water.
    b She uses 200 ml of squash.
        How much water does she
        use?
    c What *fraction* of the dilute
        drink is squash?
    d If the jug contains 1500 ml of
        the dilute drink, how much of
        the squash has Sonja used?

Exercise 7:11

Simplifying ratios

Example   Simplify these ratios. **a** 8 : 12   **b** 24 : 6 : 12

    a 4 goes into 8 and 12 $8 : 12 = \dfrac{8}{4} : \dfrac{12}{4} = 2 : 3$

    b 6 goes into 24, 6 and 12 $24 : 6 : 12 = \dfrac{24}{6} : \dfrac{6}{6} : \dfrac{12}{6} = 4 : 1 : 2$

---

1 Simplify these ratios.

    a $20:5$ **c** $10:5:20$ **e** $7:7:14$ **g** $9:3:6:3$

    b $6:4:2$ **d** $5:5$ **f** $12:6:6$ **h** $24:16:8:80$

Example   Carl, Rani and Katy win a prize of £60. They decide to share the prize in the ratio of their ages. Carl is 15, Rani is 10 and Katy is 5. How much does each of them get?

The ratio of their ages is $15:10:5 = \dfrac{15}{5}:\dfrac{10}{5}:\dfrac{5}{5} = 3:2:1$

$3 + 2 + 1 = 6$ shares are needed.

One share is £60 ÷ 6 = £10   Carl gets £10 × 3 = £30

                                Rani gets £10 × 2 = £20

                                  Katy gets £10 × 1 = £10

Check: £30 + £20 + £10 = £60

---

2 Share these amounts in the ratios given.
You may need to simplify the ratio first.
Check your answer each time.

    a £21 $2:1:4$ **c** £3 $4:4:2$ **e** £4.40 $2:8:10$

    b £27 $8:10$ **d** £34 $7:10$ **f** 70 p $18:9:3$

3 Jenny is making some lemon and blackcurrant drink. She uses three parts lemon squash, one part blackcurrant syrup and ten parts water.
Jenny has made 3.5 litres of the drink.

    a How much lemon squash has Jenny used?

    b How much blackcurrant syrup has she used?

Sometimes ratios are written in the form $1:n$ or $n:1$

Example   Convert the ratio $2:5$ into the form **a** $1:n$ **b** $n:1$

a To convert $2:5$ into the form $1:n$ you need to change the 2 to a 1. This means you divide by 2.

$$2:5 = \frac{2}{2}:\frac{5}{2} = 1:2.5$$

b To convert $2:5$ into the form $n:1$ you need to change the 5 to a 1. This means you divide by 5.

$$2:5 = \frac{2}{5}:\frac{5}{5} = 0.4:1$$

Exercise 7:12

1 Convert these ratios into the form $1 : n$.
Give your answers correct to 1 dp when you need to round.
a $4 : 5$ b $2 : 3$ c $3 : 7.5$ d $7 : 2$ e $16 : 11$

2 Convert the ratios in question **1** into the form $n : 1$.
Give your answers correct to 1 dp when you need to round.

3 Cake mixture A uses four parts fat to seven parts flour.
Cake mixture B uses five parts fat to nine parts flour.
a Write the ratio of fat to flour for each cake mixture in the form $1 : n$.
b Saleem prefers a diet with less fat and more flour.
Compare the two ratios.
Which cake mixture should Saleem choose?

Exercise 7:13

1 a Marie gets paid £20 for 5 hours work.
What does she get paid for working 1 hour?
b How much does Marie get for 7 hours work if she is paid at the same rate?

2 a Three tea bags contain 7.5 g of tea.
How much tea does one tea bag contain?
b How much tea do 10 tea bags contain?

3 4 cm^3 of copper weigh 35.68 g.
a The mass of 1 cm^3 of a material is called its **density**.
Find the density of copper in g/cm^3.
b Find the mass of 15 cm^3 of copper.

4 10 cm^3 of iron weigh 78.6 g.
a Find the density of iron in g/cm^3.
b Find the mass of 2.5 cm^3 of iron.

5 A car travels 150 miles in 3 hours.
a Write down the speed of the car in miles per hour.
b How far does the car travel in 4 hours at the same speed?

6   A plane flies 2500 km in 5 hours.
    a   Write down the speed of the plane in kilometres per hour.
    b   How far does the plane fly in 7 hours at the same speed?

7   A plane flies 1290 miles in 3 hours.
    How far does it fly in 5 hours at the same speed?

There are some formulas you can use for questions on speed, distance and time.

$$\text{Speed} = \frac{\text{Distance}}{\text{Time}} \qquad \text{Time} = \frac{\text{Distance}}{\text{Speed}} \qquad \text{Distance} = \text{Speed} \times \text{Time}$$

Here is an easy way to remember the formulas.

Look at this triangle.

$$\boxed{\begin{array}{c} D \\ S \quad T \end{array}}$$

Cover **S**	Cover **T**	Cover **D**
$S = \dfrac{D}{T}$	$T = \dfrac{D}{S}$	$D = S \times T$

Examples

1   A train travels for 210 miles at a speed of 70 miles per hour. How long does the train take?

You need to find the **time** so you use $\mathbf{T} = \dfrac{\mathbf{D}}{\mathbf{S}} = \dfrac{210}{70} = 3$ hours

2   A plane flies at 600 km per hour for 2 hours and 15 minutes. How far does it fly?
You need to find the **distance** so you use $\mathbf{D} = \mathbf{S} \times \mathbf{T}$

15 minutes is $\frac{1}{4}$ of an hour or 0.25 of an hour.
$2\frac{1}{4}$ hours = 2.25 hours
$\mathbf{D} = \mathbf{S} \times \mathbf{T} = 600 \times 2.25 = 1350$ km

If you have a [°'"] key, you can key in hours and minutes.

[6][0][0][×][2][°'"][1][5][°'"][=]

Exercise 7:14

1 A train travels for 3 hours 30 minutes at a speed of 90 miles per hour.
How far does the train travel?

2 A runner runs 1500 m in 4 minutes.
What is the runner's speed in metres per minute?

3 Kevin sees a flash of lightning.
He starts counting and counts
7 seconds until he hears the
thunder. Kevin knows that the
speed of sound is 360 metres
per second. He works out how
far the noise of the thunder
travelled to reach him.
Work out Kevin's answer in
a metres   **b** kilometres

4 In the United States, pronghorn antelopes have been timed running
4 miles in 6.8 minutes.
a Work out the speed of the antelopes in miles per minute.
Give your answer correct to 3 dp.
b Convert the speed to miles per hour.
Give your answer correct to the nearest whole number.
c You can check your answer to **b** using the [° ' ''] key.
For 6.8 minutes key in: [0] [° ' ''] [6] [.] [8] [° ' '']

5 The fastest speed recorded for a spider running across a flat surface is
53 cm per sec.
How long would the record breaking spider take to run across a room
5 m wide? Give your answer in seconds correct to 1 dp.

6 A jet plane is travelling at mach 2 (twice the speed of sound).
The plane flies 1500 km. The speed of sound is 360 metres per second.
a Work out the plane's speed in kilometres per hour.
b Work out the time the flight lasts, in hours and minutes, correct to
the nearest minute.

The same method can be used to work out **Volume**, **Density** and **Mass**.

Here is the 'density triangle'.

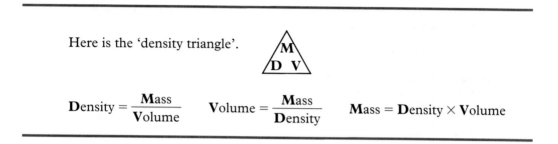

$$\textbf{Density} = \frac{\textbf{Mass}}{\textbf{Volume}} \qquad \textbf{Volume} = \frac{\textbf{Mass}}{\textbf{Density}} \qquad \textbf{Mass} = \textbf{Density} \times \textbf{Volume}$$

Exercise 7:15

1 Find the missing numbers in this table on density.
Give your answers correct to 2 sf.

Substance	Mass (g)	Volume (cm³)	Density (g/cm³)
a Aluminium	130	48	...
b Ethanol	51	...	0.789
c Silver	...	45.5	10.5
d Lead	6225	550	...
e Air	1	...	0.0012

2 This gold brooch weighs 48.25 g.
If the density of gold is 19.3 g/cm³,
find the volume of gold used to
make the brooch.

3 Robert weighs an empty flask.
He then measures 1.5 litres of salt solution into the flask and weighs it again.
The empty flask weighs 138 g. The flask plus the salt solution weighs 1723 g.
Find the density of the salt solution.
Give your answer correct to 2 dp.

4 a The density of air is 0.0012 g/cm³. Work out the mass of 1 m³ of air.
  b Find the mass of air in a classroom 6 m wide, 8 m long and 3 m high.
  Give your answer in kilograms.

1 What percentage is X of Y if:
    a X is the same size as Y   **c** X is half as big as Y
    b X is twice as big as Y   **d** X is one and a half times Y

2 This table shows the membership of a sports club.

Type of membership	Adult	Student	Child
Number of members	245	78	92

    a What is the total number of members?
    b What percentage of the membership are adults?
    c What percentage of the membership are children?
    Give your answers correct to 1 dp.

3 Michael says that 15% of 60 is the same as 60% of 15.
Is Michael correct? Show your working.

4 This marmalade has 53 g of oranges
for every 100 g of marmalade.
    a What percentage of the marmalade
       is oranges?
    b This jar contains 454 g of marmalade.
       How many grams of oranges does it contain?
       Give your answer correct to the nearest gram.

5 Brian bought a new car.
At the end of *two* years Brian had
the car valued. It was worth £6900.
Brian estimated that the value of the
car had decreased by about 15% a
year.
What did Brian pay for the car?
Give your answer correct to the
nearest £10.

6 Nathan has bought some furniture in a sale.
The prices have been reduced by $\frac{2}{5}$.
The full price of the items he bought was £390.
How much did Nathan pay?

7 Tom bought a new car in January. By September it had lost 20% of its
value. It was now valued at £5480.
How much did Tom pay for his new car?

8 Jane bought an antique table 3 years ago.
During that time its value increased by 15% to £644.
How much did Jane pay for the table?

9 The population of a village has increased by 7% over 5 years.
The new population is 1498.
What was the population 5 years ago?

10 Kevin's new house cost him £32 200. It had been reduced by 8%.
What was the full price?

11 Lisa, Jill and David win a prize of £48.
They share the prize in the ratio of their ages.
Lisa is 4 years old, Jill is 12 and David is 16.
How much does each of them get?

12 An insect repellant cream contains 15% active ingredient.
Express the proportion of active ingredient to cream base in the form:
a $1 : n$ **b** $n : 1$
Give each answer correct to 3 sf.

13 8 video cassettes put side by side
on a shelf measure 20 cm.
How many centimetres would
11 video cassettes placed side
by side measure?

14 **a** A train travels at 110 miles per hour for $2\frac{1}{2}$ hours.
How far does the train travel during this time?
b The train continues for a further half an hour at 50 miles per hour.
How far does the train travel in this time?
c Find the average speed for the whole journey.

$$\left(\text{Average speed} = \frac{\text{Total distance}}{\text{Total time}} \right)$$

15 40 cm³ of mercury weigh 544 g.
a Calculate the density of mercury.
b Calculate the mass of 68 cm³ of mercury.

1   Helen invested a sum of money in a savings account.
The interest rate was 5.5% per year.
After two years Helen's money plus the interest it had earned came to £445.21
How much did Helen invest?

2   The diagram shows three cubes of side 1 cm, 2 cm and 3 cm respectively.

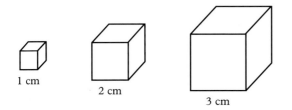

1 cm

2 cm

3 cm

a (1) Find the total length of all the edges for each of the three cubes.
     (2) Write down and simplify the ratio of the edge lengths of the cubes.
b (1) Work out the total surface area of each of the three cubes.
     (2) Write down the ratio of the surface areas of the three cubes.
c (1) Work out the volume of each of the three cubes.
     (2) Write down the ratio of the volumes of the three cubes.

3   On squared paper, draw an x axis from -5 to 11 and a y axis from -2 to 7.
Plot the points A $(-4, -1)$ and B $(10, 6)$.
Join the points to get the straight line AB.
Point C divides AB in the ratio $5 : 2$. Find the co-ordinates of C.

4   The mean density of the earth is 5.52 g/cm^3.
The volume of the earth is 1.08×10^{12} km^3.
Calculate the mass of the earth.
Give your answer in standard form correct to 2 sf.

5   The earth is an average distance of 93 000 000 miles from the sun.
a Light travels at 186 000 miles per second.
    How long does it take light to reach the earth from the sun?
    Give your answer correct to the nearest minute.
b It takes a year for the earth to go round the sun.
    Take the earth's orbit round the sun as an approximate circle.
    Calculate the speed of the earth relative to the sun in miles per hour.
    Give your answer correct to 2 sf.

6   A metal worker is using an alloy made of equal volumes of silver and another
metal. 35.2% by weight of the alloy is silver.
The density of silver is 10.5 g/cm^3.
Find the density of the other metal correct to 1 dp.

• *Examples* **1** Change these to percentages: **a** 0.65 **b** $\frac{3}{5}$

$\boxed{\times 100}$

a $0.65 \times 100 = 65\%$

b $\frac{3}{5} \times 100 = 60\%$

2 Change 14% to: **a** a decimal **b** a fraction in its lowest terms

$\boxed{\div 100}$

a $14\% = 14 \div 100 = 0.14$ **b** $14\% = \frac{14}{100} \xrightarrow{\div 2} \frac{7}{50}$ $\xleftarrow{\div 2}$

• *Example* A coat is reduced by 15% in a sale.
Rachel buys the coat for £61.20.
Find the full price of the coat.

The *full* price is 100%.
The full price is reduced by 15%
to get the sale price.
The *sale* price is $100\% - 15\% = 85\%$

85% of the full price is £61.20

1% is $\dfrac{£61.20}{85} = £0.72$

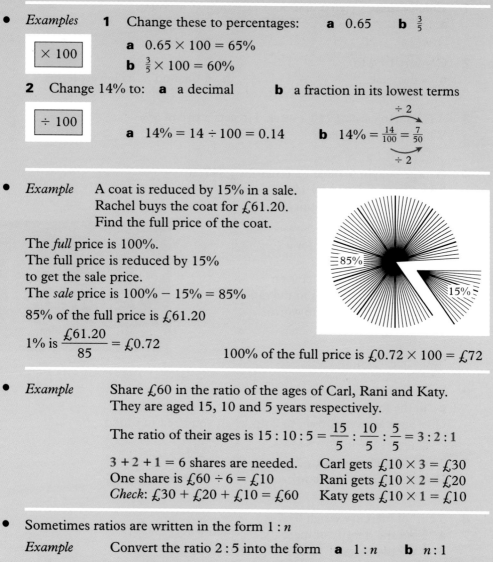

100% of the full price is $£0.72 \times 100 = £72$

• *Example* Share £60 in the ratio of the ages of Carl, Rani and Katy.
They are aged 15, 10 and 5 years respectively.

The ratio of their ages is $15:10:5 = \dfrac{15}{5} : \dfrac{10}{5} : \dfrac{5}{5} = 3:2:1$

$3 + 2 + 1 = 6$ shares are needed. Carl gets $£10 \times 3 = £30$
One share is $£60 \div 6 = £10$ Rani gets $£10 \times 2 = £20$
Check: $£30 + £20 + £10 = £60$ Katy gets $£10 \times 1 = £10$

• Sometimes ratios are written in the form $1:n$

Example Convert the ratio $2:5$ into the form **a** $1:n$ **b** $n:1$

a To convert $2:5$ into the form $1:n$ **b** To convert $2:5$ into the form $n:1$
you need to change the 2 to a 1. you need to change the 5 to a 1.

$2:5 = \dfrac{2}{2} : \dfrac{5}{2} = 1:2.5$ $2:5 = \dfrac{2}{5} : \dfrac{5}{5} = 0.4:1$

• Here is an easy way to remember the formulas for **Speed**, **Distance** and **Time**.

Look at this triangle.

$\triangle \begin{smallmatrix} D \\ S \ T \end{smallmatrix}$

Cover **T** to get $T = \dfrac{D}{S}$

Cover **S** to get $S = \dfrac{D}{T}$

Cover **D** to get $D = S \times T$

1   Change these to percentages.
    a $\frac{4}{5}$ **b**   0.39

2   Convert 16% to:
    a   a decimal   **b**   a fraction in its lowest terms.

3   A salesman earns £240 a week. He gets a rise of 5%.
    What is his new salary?

4   George bought a video recorder for £367.50
    It had been reduced by $12\frac{1}{2}\%$.
    What was the original price?

5   The total prize money for a
    marathon run is £360.
    The money is shared between
    1st, 2nd and 3rd prizes in the ratio 5 : 3 : 1.
    How much is each prize worth?

6   Express the ratio 5 : 12
    a   in the form 1 : n **b**   in the form n : 1
    Give your answer correct to 2 dp when you need to round.

7   Three model cars cost £2.61
    How much do 7 cars cost?

8   5 cm³ of mercury weigh 67.7 g.
    a   Find the density of mercury.
    b   Find the mass of 12 cm³ of mercury.

9   A plane travels 2320 km at a speed of 580 km per hour.
    How long does the plane take to fly this distance?

10 The speed of light is about 300 000 km per second.
    How far does light travel in 6 seconds?

11 A porcelain vase weighs 563.5 g.
    If the density of the porcelain is 2.3 g per cm³, find the volume of
    porcelain used to make the vase.

8 Algebra

In 3 years' time, Nishi's father will be 4 times as old as Nishi.

In 3 years' time, Nishi's father will be 40.

How old is Nishi now?

1 Brackets

... *continued* on page 159 ...

◄◄REPLAY►

Collecting terms	$a + a + a + a + a = 5a$ This is called **collecting terms**. *Remember*: $5a = 5 \times a$
Power	$a \times a \times a = a^3$ The **power** '3' tells you how many as are multiplied together.

Exercise 8:1

Write each of these expressions in a shorter form by collecting terms or using a power.

1 $b + b + b + b + b + b$

2 $m \times m$

3 $g + g + g + g$

4 $y \times y \times y \times y \times y$

5 $t + t + t$

6 $k + k$

7 $r \times r \times r \times r$

8 $j \times j \times j \times j \times j \times j \times j$

9 $p + p + p + p + p$

10 $h \times h \times h$

You can use a number ladder to help you if some of the numbers are negative.

Examples   Use the number ladder to help you with these.

Start at 0.
Count up for positive numbers.
Count down for negative numbers.

1 $3 - 6 + 4 = 1$

2 $4r - 5r - 2r = -3r$

```
 7
 6
 5
 4
 3
 2
 1
 0
-1
-2
-3
-4
-5
-6
-7
```

Exercise 8:2

Collect these terms.

1 $3 - 1 - 4$

2 $-5 + 4 - 3$

3 $3h + 2h - 6h$

4 $5f - 2f + f$

5 $3s + 3s - 8s$

6 $7y - 5y - 2y$

7 $6k - 5k - 4k$

8 $3d - d + 2d - 5d$

9 $7r - 4r - 5r - 2r$

10 $5p - p + 3p - 2p$

Sometimes you have more than one letter or number to collect.

Examples   **1** $f + f + f + g + g = 3f + 2g$

2 $2a - 4a + 3b = -2a + 3b$

3 $4t - 2 + 3t = 7t - 2$

4 $3g - g + 2h - 4h = 2g - 2h$

Exercise 8:3

Collect like terms.

1 $j + j + k + k + k + k$ **7** $4g - 8g + 5h + 2h$

2 $p + p + q - q + q$ **8** $2p - 4p + 3q - 3q$

3 $h + h - h - h + i$ **9** $3a + b + a - 2b$

4 $5f - 2f + g - g$ **10** $7r + 2s - r - s$

5 $r - 2r + 3s + 2s$ **11** $5p + 6 - 7p - 9$

6 $7y - 5y - 3y + 4$ **12** $3t + 6s - 8s + t$

Sometimes you have more than one letter or number to multiply.

Examples   **1** $f \times g = fg$ **3** $(2p)^2 = 2p \times 2p = 4p^2$

      2 $g \times 3h = 3gh$ **4** $4r \times 2rs = 8r^2s$

Exercise 8:4

Simplify these expressions.

1 $2 \times 4c$ **4** $3s \times 3t$ **7** $3a \times 2a$ **10** $3ik \times 6k$

2 $2 \times r \times s$ **5** $4b \times 5c$ **8** $(4k)^2$ **11** $p \times 3pq$

3 $5 \times 3c$ **6** $2e \times 6e$ **9** $(2q)^2$ **12** $4r \times 2rs$

Different combinations of letters are collected separately.

Examples   **1** $2f + 3fg - 2gf + 4g = 2f + fg + 4g$ (*fg* is the same as *gf*)

      2 $2x + 3x^2 + 4x + x^2 = 6x + 4x^2$

Exercise 8:5

Simplify these expressions.

1 $5ab - 3ab$

7 $4q + 5pq - 6q - 6qp$

2 $3x^2 - 4x^2$

8 $4y + 3y^2 - 7y^2 - 2y$

3 $6vw - 4w + 5wv$

9 $xy + x^2 - 3xy + 3x^2$

4 $5c + 3cd - 5cd$

10 $5pq - 4q - 3q + 2qp$

5 $7y - xy + 3xy + 2x$

11 $8r + 6rs - 2sr - 3r$

6 $x + 2x - 3x^2 + 5x^2$

12 $3s + 2st - 4ts - 4t$

Multiplying out brackets

You can use BODMAS to remind you what to do first.

You do	**Brackets** first
then powers	**Of**
Next you do	**Division**
and	**Multiplication**
Then you do	**Addition**
and	**Subtraction**

Example Work these out using the rules of **BODMAS**.

a $3 \times (5 + 9)$

b $2 \times 4 + 3 \times 5$

$$\begin{aligned}\textbf{a} \quad 3 \times (5 + 9) &= 3 \times 14 \\ &= 42\end{aligned}$$

$$\begin{aligned}\textbf{b} \quad 2 \times 4 + 3 \times 5 &= 8 + 15 \\ &= 23\end{aligned}$$

Exercise 8:6

1 Work out part (1) and part (2) of **a** to **d** using the rules of BODMAS.

a (1) $2 \times (3 + 7)$ (2) $2 \times 3 + 2 \times 7$

b (1) $5 \times (9 - 4)$ (2) $5 \times 9 - 5 \times 4$

c (1) $10 \times (7 + 8)$ (2) $10 \times 7 + 10 \times 8$

d (1) $7 \times (6 - 5)$ (2) $7 \times 6 - 7 \times 5$

2 Write down what you notice about the answers in question **1**.

If you have a letter in the bracket, you cannot work out the bracket first. You have to *multiply out* the bracket.

Using the rules of algebra $2(3j + 4)$ means $2 \times (3j + 4)$

$2 \times (3j + 4)$ means $2 \times 3j + 2 \times 4 = 6j + 8$

Example   Multiply out the bracket $7(10 - r)$

$$7(10 - r) = 7 \times 10 - 7 \times r$$
$$= 70 - 7r$$

---

3 Copy these expressions and fill them in.

a $4 \times (3 + a) = 4 \times 3 + 4 \times \ldots$ **c** $5 \times (2p - 3) = 5 \times \ldots - 5 \times \ldots$
   $= 12 + \ldots$ $= \ldots - \ldots$

b $7 \times (t - 3) = 7 \times \ldots - 7 \times \ldots$ **d** $10 \times (2 + 3h) = 10 \times \ldots + 10 \times \ldots$
   $= 7t - \ldots$ $= \ldots + \ldots$

Exercise 8:7

Multiply out these brackets.

1 $4(3 + t)$ **4** $9(2s - 3)$ **7** $6(5 + j + k)$

2 $6(1 - s)$ **5** $4(2t + s)$ **8** $8(2 + m + n)$

3 $4(p + q)$ **6** $3(10j - 4k)$ **9** $9(r - s - t)$

You can have letters outside the bracket.
Remember: $d \times 5 = 5d$ You would **not** write $d5$.

Example   Multiply out: **a** $d(5 + c)$ **b** $f^2(2f + 3)$

a $d(5 + c) = d \times 5 + d \times c$ **b** $f^2(2f + 3) = f^2 \times 2f + f^2 \times 3$
   $= 5d + cd$ $= 2f^3 + 3f^2$

---

10 $a(5 + b)$ **13** $c^2(c + 2)$ **16** $5t^2(s + t)$

11 $2f(g - 3)$ **14** $e^2(5 - e)$ **17** $2x^3(x - y)$

12 $3m(2n + 5)$ **15** $r^2(3 - 2s)$ **18** $3r(2r - 3s - t)$

Factorising

To **factorise** an expression in algebra you *put brackets in.*

Example

Factorise: **a** $6c + 3$ **b** $18 - 12f$

a   3 is a factor of 6 and 3 $6c + 3 = 3(2c + 1)$

Notice that you need a 1 at the end of the bracket.

b   6 is a factor of 18 and 12 $18 - 12f = 6(3 - 2f)$

For part **b**, $2(9 - 6f)$ or $3(6 - 4f)$ would not be completely factorised.
The terms in the brackets still have a common factor.
You need to take out the *biggest* factor possible.

Check your factorising by multiplying out the brackets again in your head.

Exercise 8:8

Factorise these.

1 $4k + 2 = 2($ $)$ **8** $24q + 36$ **15** $2h + 4i + 4j$

2 $3f + 3 = 3($ $)$ **9** $27h - 18$ **16** $6y - 6z - 6$

3 $5 + 10p = 5($ $)$ **10** $18 + 6a$ **17** $7c + 14d - 14e$

4 $7 - 14k = 7($ $)$ **11** $10j + 25$ **18** $9g + 3h - 6i$

5 $8j + 4 = 4($ $)$ **12** $8 - 24r$ **19** $8i - 4j - 8k$

6 $18y - 9 = 9($ $)$ **13** $3r + 3s + 3t$ **20** $6a + 12b - 6$

7 $15 + 20t = 5($ $)$ **14** $4p + 4q + 4$ **21** $12e + 4f - 6g$

Example   Factorise: **a** $2y^2 + y$ **b** $2c^2 + 3bc$ **c** $16k^2 - 24k^3$

a y is a factor of y^2 and y

$2y^2 + y = y(2y + 1)$
(Notice that you need a 1 at the end of the bracket.)

b c is a factor of c^2 and bc

$2c^2 + 3bc = c(2c + 3b)$

c 8 is a factor of 16 and 24
k^2 is a factor of k^3 and k^2

$16k^2 - 24k^3 = 8k^2(2 - 3k)$

Exercise 8:9

Factorise these.

1 $3t + 4t^2 = t (\quad)$

2 $5h - 6h^2 = h (\quad)$

3 $7xy - y = y(\quad)$

4 $x^2 + x$

5 $7g + gh$

6 $pq - p^2$

7 $3f^2 + 6f = 3f(\quad)$

8 $4r + 8rs = 4r(\quad)$

9 $9jk - 36j = 9j(\quad)$

10 $24p^2 + 30pq$

11 $15t^2 - 21t^3$

12 $6t^3 - 18t^2$

13 $6j^3 + 3j^2$

14 $5r^3 + 10r^2 - 15r$

15 $4t^3 + 8t^2 - 8t$

16 $3s - 3s^2 - 6s^3$

17 $8d - 12d^2 - 20d^3$

18 $16h - 8gh - 12gh^2$

19 $3y^2 + 3xy + 3x^2y$

20 $4b^3 - 4b^2 - 4b$

21 $2g + 4g^2 + 2g^3$

22 $7rs^2 - 14r^2 - 14rs$

● **23** $12x^2y + 6xy^2$

● **24** $30ab^2 + 36a^2b^2$

2 Two brackets

... *continued* on page 164 ...

◀◀REPLAY▶

Rules for multiplying directed numbers:

$\left. \begin{array}{c} + \times + \\ - \times - \end{array} \right\}$ makes a + $\left. \begin{array}{c} + \times - \\ - \times + \end{array} \right\}$ makes a −

Examples

1 $-3 \times 4 = -12$

3 $7 \times -2y = -14y$

2 $-8 \times -5 = 40$

4 $-2g \times -g = 2g^2$

Exercise 8:10

Multiply these.

1 4×-5

4 $-h \times -2h$

7 $-p \times -6p$

10 $t \times 3t^2$

2 -2×-3

5 $3c \times -4$

8 $-6 \times -2w$

11 $-4r \times -5r$

3 5×-2

6 $-k \times -4k$

9 $-2t \times -3s$

12 $-y^2 \times 2y$

Example Multiply out the brackets: **a** $-5(y + 7)$ **b** $-(4 - t)$

a $-5(y + 7) = -5 \times y + -5 \times 7$
$= -5y - 35$

b $-(4 - t) = -4 - -t$
$= -4 + t$

Exercise 8:11

Multiply out these brackets.

1 $-2(3 + g)$ **4** $-5(r - s)$ **7** $-p(p - 4)$ **10** $-3h(5 - h)$

2 $-(x + y)$ **5** $-(r - 5)$ **8** $-2g(g - h)$ **11** $-5s(s + 4)$

3 $-3(r + 1)$ **6** $-7(s - t)$ **9** $-(4r - 3)$ **12** $-9y(y - 1)$

You can have more than one bracket.

Example   Multiply out these brackets.
Give each answer in its simplest form.
a $5(3 + 2x) - 3(4 - x)$ **b** $a(a + 4) + 3(2a - 1)$

a $5(3 + 2x) - 3(4 - x) = 15 + 10x - 12 + 3x$
$$= 3 + 13x$$

b $a(a + 4) + 3(2a - 1) = a^2 + 4a + 6a - 3$
$$= a^2 + 10a - 3$$

Exercise 8:12

Multiply out these sets of brackets.
Give each answer in its simplest form.

1 $2(3 + y) + 5(4 + y)$ **9** $2(5r + 9) - 6(r - 2)$

2 $3(4 + d) + 4(2 + d)$ **10** $s(s + 1) + 5(s + 1)$

3 $6(3 + x) + 5(2 - x)$ **11** $x(2x + 1) + 2(3x + 4)$

4 $3(f + 8) + (f + 4)$ **12** $2d(3d + 1) + 5(d + 2)$

5 $4(3k + 2) + 2(4k - 1)$ **13** $2s(s + 2) + 3(2s + 3)$

6 $2(10 + 5e) - 3(6 + e)$ **14** $y(y - 2) + 6(y + 1)$

7 $3(4r + 1) - (7r - 2)$ **15** $3c(c + 4) - 2(3c + 2)$

8 $4(w + 1) - (w - 1)$ **16** $4(5 + 2f) + f(3 + f)$

17 Sally's rectangle is a cm long and $a - 5$ cm wide. Sally cuts a piece 3 cm wide off the rectangle, as shown in the diagram.
Sally writes down a formula for the area of the piece that is left.
$a(a - 5) - 3(a - 5)$

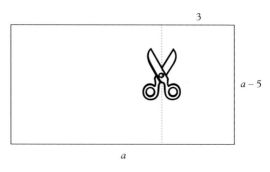

a Multiply out the brackets and simplify Sally's formula.

b Find the area of the piece that is left when $a = 7.5$ cm.

Multiplying a bracket by a bracket

Example Multiply out: $(s + 3)(2s - 5)$

You split the first bracket up.
Then you multiply the second bracket by both the s and the 3.

$$(s + 3)(2s - 5) = s(2s - 5) + 3(2s - 5)$$
$$= 2s^2 - 5s + 6s - 15$$
$$= 2s^2 + s - 15$$

Exercise 8:13

Multiply out these brackets.

1 $(y + 2)(y - 4)$

2 $(s + 1)(3s + 2)$

3 $(2 + f)(1 + 4f)$

4 $(d - 2)(3d + 5)$

5 $(7 + k)(1 + k)$

6 $(a + 3)(4a - 1)$

7 $(y - 2)(y + 2)$

8 $(a + b)(a - b)$

9 $(j + 2)^2 = (j + 2)(j + 2)$

10 $(3s - t)^2$

Here are two ways to help you multiply out brackets.

Examples **1** Multiply out these brackets: $(y + 4)(2y - 6)$

Cover up the 4 and multiply out the bracket using the y.

$(y \quad\quad (2y - 6) \quad = \quad 2y^2 - 6y \ldots$

Cover up the y and multiply out the bracket using the 4.

$+ 4)(2y - 6) \quad = \quad \ldots 8y - 24$

$2y^2 - 6y + 8y - 24 = \quad 2y^2 + 2y - 24$

2 Multiply out: $(2x - 5)(x + 3)$

This 'face' helps some people multiply out brackets.

$2x^2 \quad -15$

$(2x - 5)\ (x + 3)$

$-5x$

$6x$

$= 2x^2 - 5x + 6x - 15$

Answer: $2x^2 + x - 15$

Try the methods in the examples when you work these out.

11 $(g + 2)(3g + 4)$ **17** $(k - 6)(3k - 8)$ **23** $(4g + h)^2$

12 $(h - 5)(2h + 4)$ **18** $(h + 3)(3h - 4)$ **24** $(8 - 5r)(1 - r)$

13 $(y + 2)(3y - 4)$ **19** $(3c + 4)(2c - 7)$ **25** $(3 + 4h)(3 - 4h)$

14 $(5 - f)^2$ **20** $(1 + 7k)(5 - k)$ **26** $(2y - 7z)(5y - 4z)$

15 $(r - s)(2r - 3s)$ **21** $(4d - 5)(d - 1)$ **27** $(5f + 3g)(f - 7g)$

16 $(1 + 2x)(4 + 3x)$ **22** $(3s - 5t)(4s - 2t)$ **28** $(2d - 3c)(2d + 3c)$

Here is a quick way of squaring a bracket.

$$(\text{first term})^2 + 2 \times \text{first term} \times \text{second term} + (\text{second term})^2$$

Example   Multiply out: **a** $(s + 3t)^2$ **b** $(5a - b)^2$

(Do this line in your head)   **a** $(s + 3t)^2 = s^2 + 2 \times s \times 3t + (3t)^2$
$$= s^2 + 6t + 9t^2$$

(Do this line in your head)   **b** $(5a - b)^2 = (5a)^2 + 2 \times 5a \times -b + b^2$
$$= 25a^2 - 10ab + b^2$$

---

29 **a** $(t + 7)^2$ **c** $(3 + y)^2$ **e** $(a + 2b)^2$ **g** $(2e + 3f)^2$
  b $(d + 5)^2$ **d** $(7 + x)^2$ **f** $(c + 10d)^2$ **h** $(3p + 4q)^2$

30 **a** $(p - 5)^2$ **c** $(x - y)^2$ **e** $(d - 2e)^2$ **g** $(3p - 5q)^2$
  b $(g - 4)^2$ **d** $(3x - y)^2$ **f** $(5m - n)^2$ ● **h** $(x^2 - y^2)^2$

31 **a** Multiply out these brackets.
       (1) $(a + 5)(a - 5)$ (2) $(4 + r)(4 - r)$ (3) $(6 - 5q)(6 + 5q)$
   b Write down what you notice.
   c Copy this formula: $(a + b)(a - b) = a^2 - b^2$
   d Write down the answers to these using the formula.
       (1) $(7 + g)(7 - g)$ (2) $(5t + 3)(5t - 3)$ (3) $(g - 2h)(g + 2h)$

32 Kiran is using coloured tiles to make this pattern.
   He wants to find a formula for the total number of tiles.

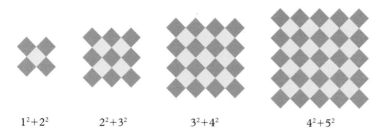

   $1^2 + 2^2$ $2^2 + 3^2$ $3^2 + 4^2$ $4^2 + 5^2$

   a Complete Kiran's formula for the total number of tiles in the *n*th shape:

   $$n^2 + (n \ ...)^2$$

   b Multiply out the bracket and simplify the formula.

3   Equations and formulas

Linear equations	Equations with simple letters and numbers in are called **linear equations**. Linear equations must not have any terms with powers like x^2 or x^3
	Remember: When you are solving equations you *must* do the same thing to both sides of the equation.

◄◄REPLAY►

Exercise 8:14

Solve these equations.

1 $4x = 16$

2 $3x = 21$

3 $x + 6 = 12$

4 $x + 12 = 45$

5 $x - 7 = 23$

6 $\dfrac{x}{5} = 12$

7 $\dfrac{x}{7} = -2$

8 $0.5x = 3$

9 $3x + 1 = -14$

10 $5x - 3 = 27$

11 $3x + 7 = 4$

12 $\dfrac{x}{5} - 1 = 12$

13 $\dfrac{x}{3} + 2 = 10$

14 $4x - 6 = 16$

15 $\dfrac{2x}{3} = -6$

16 $\dfrac{3x}{2} + 5 = 26$

Look at the equation $7x = 4x + 9$
It has letters on both sides.
To solve it, you start by changing the equation so that x is only on one side.

Example   Solve $7x = 4x + 9$

Look to see which side has the *least x*.
In this example the right-hand side has only $4x$.
Subtract $4x$ from each side. $7x - 4x = 4x - 4x + 9$
$$3x = 9$$
Divide both sides by 3. $x = 3$

Exercise 8:15

Solve these equations.

1 $6x = 4x + 8$

2 $9x = 6x + 24$

3 $10x = 5x + 45$

4 $7x = x + 42$

● **5** $4x - 8 = 6x$

6 $12x = 6 + 10x$

7 $2.5x = 1.5x + 6$

● **8** $3.5x = x + 10$

Sometimes, the side with the least x will be the left-hand side.
You can still solve the equation in the same way.

Example   Solve $3x + 6 = 6x$

Take $3x$ from both sides. $3x + 6 - 3x = 6x - 3x$
$$6 = 3x$$
Divide by 3. $2 = x$

It is usual to write this
the other way around. $x = 2$

9 $4x + 3 = 5x$

10 $3x + 6 = 5x$

11 $2x + 17 = 4x$

● **12** $5x - 21 = 12x$

13 $0.5x + 15 = 2.5x$

14 $3x - 6 = 5x$

Some equations have letters and numbers on both sides.
First, change the equation so it has x on only one side.

Example $5x + 3 = 2x + 15$

The right-hand side has least x.
Subtract $2x$ from each side. $5x - 2x + 3 = 2x - 2x + 15$
$$3x + 3 = 15$$

Now remove the numbers from the side with the x.
The left-hand side has a 3.
Subtract 3 from each side. $3x + 3 - 3 = 15 - 3$
$$3x = 12$$
Divide both side by 3. $x = 4$

Exercise 8:16

Solve these equations.

1 $6x + 2 = 3x + 5$

2 $9x + 1 = 5x + 13$

3 $5x - 5 = 2x + 4$

4 $10x - 12 = 3x + 16$

5 $13x - 15 = 12x + 19$

6 $8x + 12 = 3x + 57$

7 $12x - 1 = 7x + 19$

8 $5x - 4 = 4x + 3$

9 $2.5x + 10 = 1.5x + 17$

10 $3.5x - 12 = x + 3$

The method does not alter if you have a minus sign in front of one of the letters.
This means that the side with the minus sign has the least x.
To remove the x from this side, **add** the same number of xs on to each side.

Example   Solve $5x + 3 = 21 - x$

First, remove the x from the right-hand side.
To do this, add x to each side. $5x + x + 3 = 21 - x + x$
Now solve the equation as before. $6x + 3 = 21$
Subtract 3 from each side. $6x = 18$
Divide both sides by 6. $x = 3$

11 $5x + 3 = 15 - x$ **15** $10x - 5 = 31 - 2x$

12 $7x + 2 = 20 - 2x$ **16** $2x - 7 = 3 - 8x$

13 $2x + 4 = 13 - x$ **17** $2.6x + 10 = 23 - 1.3x$

14 $4x - 3 = 12 - x$ **18** $0.5x - 5 = 15 - x$

If equations have brackets in them, you should remove the brackets first.

Example   Solve $3(2x + 1) = 27$

Remove the bracket.

$$3 \times 2x + 3 \times 1 = 27$$
$$6x + 3 = 27$$

Now solve the equation as usual.

$$6x = 24$$
$$x = 4$$

Exercise 8:17

Solve these equations.

1 $3(2x + 2) = 24$ **11** $6(3x + 1) = 3(4x + 2)$

2 $5(6x - 2) = 50$ **12** $2(7x - 3) = 3(2x + 6)$

3 $4(3x - 7) = 20$ **13** $3(6x - 5) = 5(x + 10)$

4 $6(7x - 1) = 36$ **14** $4(4x + 3) = 5(5x + 2)$

5 $10(3x - 4) = -80$ **15** $6(3x + 2) = 3(4x + 6)$

6 $12(3x + 2) = 36$ **16** $10(3x - 4) = 5(6 - x)$

7 $2(x + 1) = -8$ **17** $7(2x - 5) + 4(6 - x) = 19$

8 $6(3x + 5) = 30$ • **18** $4(7 + 3x) - 5(2 + x) = 46$

9 $7(2x - 16) = 0$ • **19** $11(3 + 2x) = 5(x + 6.5)$

10 $\frac{1}{2}(x + 5) = 20$ • **20** $4(4x + 3) + 6(2 - x) - 3(2x - 1) = 19$

Changing the subject

Changing the subject of a formula uses the same skills as solving an equation.

Instead of getting a number as an answer, you are trying to get a single letter on one side of the formula.

This single letter is known as the **subject** of the formula.

For example, in the formula $v = u + at$, v is the subject.

Examples

1 Make y the subject of the formula $\quad x = y - 2z$

We need to remove the $2z$ from the right-hand side to leave the y by itself.

$$x = y - 2z$$

Add $2z$ to each side. $\qquad\qquad x + 2z = y$

This can then be written as $\qquad y = x + 2z$

2 Make t the subject of the formula $\quad c = \dfrac{zt + zb}{2}$

Look at the letter t on the right-hand side of the formula.

In order it has: been multiplied by z
had zb added to it
been divided by 2

These are the operations that need undoing by using their inverses.

First **multiply** by 2 $\qquad 2c = zt + zb$

Subtract zb $\qquad\qquad 2c - zb = zt$

Divide by z $\qquad\qquad \dfrac{2c - zb}{z} = t$

This can be rewritten as $\qquad t = \dfrac{2c - zb}{z}$

Exercise 8:18

Change the subject of the these formulas.
The new subject should be the letter in **red**.

1 $c = a - b$

2 $g = 5f + 3$

3 $t = sk - h$

4 $h = \dfrac{y}{4}$

5 $p = \dfrac{t - 5z}{7}$

6 $v = u + at$

7 $m = k + nk$

8 $p = \dfrac{5t - 6z}{7}$

9 $y = \dfrac{tx + rt}{2}$

10 $f = st - u^2$

11 $y + 5 = tm + 6t$

12 $t^2 = 7r + \dfrac{x}{4}$

Some formulas have squares and square roots in them.
Squares and square roots are the inverses of each other.
To remove a square, take the square root of each side.
To remove a square root, square each side.

Example   Make t the subject of the formula $s = \sqrt{t + r}$

To remove the square root, square each side:

$$s^2 = t + r$$

Now subtract the r $s^2 - r = t$

This can be written as $t = s^2 - r$

---

13 $s = \sqrt{2t - r}$

14 $d = \sqrt{3t + 6y}$

15 $p = \sqrt{\dfrac{q}{7}}$

16 $A = \pi r^2$

17 $t = s^2 + 5$

18 $t^2 = s^2 - 6f$

19 $s = \frac{1}{2}gt^2$

20 $y = \sqrt{\dfrac{6t}{7}}$

21 $v^2 = u^2 + 2as$

22 $T = 2\pi\sqrt{\dfrac{l}{g}}$

● **23** $h = \sqrt{g^2 - 5}$

● **24** $y - 5 = fg - r^2$

1 Simplify these expressions.
   a -6×7 **c** $-4j \times -jk$
   b $-2h^2 \times -5h$ **d** $3t \times -5t^2$

2 Simplify these expressions.
   a $5h + 3g - 4h - 7g$ **c** $d - 8c - 5c + 3d$
   b $6s - 5 + s + 4$ **d** $6p + 3q - 7p - 8p$

3 Factorise:
   a $6t + 12s$ **c** $5t^2 + 15t$ **e** $rs - s^2$ **g** $16z^2 - 8yz$
   b $d^2 - d$ **d** $xy - 5y$ **f** $7g + g^2h$ **h** $30p^2q + 24pq^2$

4 Multiply out these brackets.
   a $3(2r + 7) + 3(3r - 5)$ **d** $5p(6 + p) - 4p(2p - 1)$
   b $4(5t - 6) - 7(t + 1)$ **e** $6(3h + 4i) - 7(2h - i)$
   c $3(k - 5) + k(k + 4)$ **f** $7(4 - 2e) - (3 + e)$

5 Multiply out these brackets.
   a $(g + 5)(2g + 3)$ **d** $(7y + 2)(y - 4)$
   b $(h + 1)(2h - 3)$ **e** $(r + 2s)(r - s)$
   c $(r - 4)(2r - 7)$ **f** $(s - 3t)(2s - 5)$

6 **a** Copy and complete: $(a + b)(a - b) = \ldots - \ldots$
     b Use the formula to simplify these.
        (1) $(s + 2t)(s - 2t)$ (2) $(f + 5g)(f - 5g)$ (3) $(p - 3q)(p + 3q)$

7 Multiply out these brackets.
   a $(r + s)^2$ **c** $(2a - b)^2$ **e** $(1 - 4e)^2$ **g** $(3a + 4b)^2$
   b $(p - q)^2$ **d** $(s + 3t)^2$ **f** $(g + 5h)^2$ **h** $(6x - 7y)^2$

8 Solve these equations.
   a $5x + 3 = 2x + 18$ **c** $4x - 5 = x - 1$
   b $12x - 4 = 7x - 29$ **d** $7x + 1 = 5x + 9$

9 Solve these equations.
   a $7x = 3x + 24$ **c** $6x + 16 = 10x$
   b $2x - 16 = 6x$ **d** $4.5x = 2x + 20$

10 Solve these equations.
   a $5(x + 3) = 25$ **c** $3(3x - 2) = 30$
   b $6(2x + 2) = 24$ **d** $5(2x + 1) = 35$

11 Petra wants to work out $(r + 4)(2r + 5)$

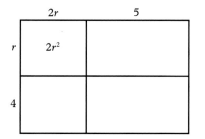

a Multiply out Petra's brackets.
b Petra uses the area of this rectangle to check her working.
 Copy Petra's rectangle.
 Fill in the missing areas inside the rectangle.
c Add up and simplify the area of the rectangle.
 Use the answer to check part **a**.

12 Four people each have a card.

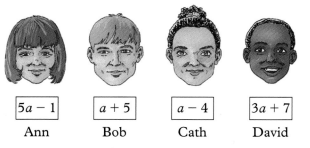

$5a - 1$	$a + 5$	$a - 4$	$3a + 7$
Ann	Bob	Cath	David

a Ann and David decide to find the value of a that makes their cards
 have the same value. They write this equation:
  $5a - 1 = 3a + 7$
 Find the value of a that makes Ann and David's cards the same.
b Ann and Bob decide to find the value of a that makes their cards
 have the same value. Find this value of a.
c There are two people whose cards will never be worth the same as
 each other.
 Who are the two people?
 Explain why their cards cannot be the same.

13 Change the subjects of these formulas.
 For each one, the new subject should be the letter in **red**.

a $s = 3r - 5p$ **c** $h = 4g^2 - 5j$ **e** $w = \dfrac{4k - m}{5}$

b $A = 6l^2$ **d** $p = \sqrt{\dfrac{q}{5}}$ **f** $v^2 = u^2 + 2as$

1   Factorise these:

    a $3x^2y - 6xy^2$ **b** $18pq^2r + 24pqr^2$ **c** $12w^3 + 24w^2 - 6w$

2   The triangle numbers are:

    1   3   6   10

    a   Sian has found a formula for the nth triangle number: $\dfrac{n(n + 1)}{2}$

       Find the fifth triangle number using Sian's formula.

       Draw the fifth triangle number to see if you are right.

    b   Sian has put some triangles together to make squares like this.

      2^2 3^2 4^2

      Sian's formula for the nth square is $\dfrac{n(n + 1)}{2} + \dfrac{(n + 1)(n + 2)}{2}$

      (1) Simplify Sian's formula to show that it is correct.

      (2) Show that $(n + 1)^2$ is the same as Sian's formula.

3   Solve the following equations.

    a $3x - (5x - 3) = x$

    b $5(3x - 3) - 2(6 - 3x) = 13x - 3$

    c $(2x - 3)(x - 4) = (2x + 1)(x + 3)$

4   Solve the following equations.

    a $\dfrac{8}{x - 3} = 4$ **b** $\dfrac{30}{2x - 3} = 6$

5   Julie is making a clockface.
Julie's clockface is a square
of area 81 cm².
Saleem wants to make a circular
clockface that has the same
area as Julie's.
Find the radius that Saleem
should use for his clock.

- Sometimes you have more than one letter or number to collect.

  Examples **1** $2a - 4a + 3b = -2a + 3b$ **2** $3g - g + 2h - 4h = 2g - 2h$

- *Example* Multiply out these brackets.

  a $2(3j + 4) = 2 \times 3j + 2 \times 4$ **b** $-(4 - t) = -4 - -t$
  $= 6j + 8$ $= -4 + t$

- To **factorise** an expression in algebra you put in brackets.

  Example Factorise: $18 - 12f$

  6 is a factor of 18 and 12 $18 - 12f = 6(3 - 2f)$

  $2(9 - 6f)$ or $3(6 - 4f)$ would not be completely factorised.
  The terms in the brackets still have a common factor.
  You need to take out the biggest factor possible.

- *Example* Multiply out: $(s + 3)(2s - 5)$

  You split the first bracket up.
  Then you multiply the second bracket by both the s and the 3.

  $(s + 3)(2s - 5) = s(2s - 5) + 3(2s - 5)$
  $= 2s^2 - 5s + 6s - 15$
  $= 2s^2 + s - 15$

- *Examples* Solve these equations.

  1 $3(2x + 1) = 27$ **2** $3x + 6 = 6x$
  Remove the bracket. Take $3x$ from both sides.
  $3 \times 2x + 3 \times 1 = 27$ $6 = 3x$
  $6x + 3 = 27$ Divide by 3. $2 = x$
  $6x = 24$ Write this the other way round.
  $x = 4$ $x = 2$

- Make t the subject of: **1** $c = \dfrac{zt + zb}{2}$ **2** $s = \sqrt{t + r}$

  First **multiply** by 2 $2c = zt + zb$ To remove the square root, square each side.
  Subtract zb $2c - zb = zt$ $s^2 = t + r$
  Divide by z $\dfrac{2c - zb}{z} = t$ Now subtract the r.
  $s^2 - r = t$
  This can be rewritten as $t = \dfrac{2c - zb}{z}$ This can be written as $t = s^2 - r$

1 Simplify these expressions.
  a $f - 5f + 4f$ **b** $4k - 5 - 8 + k$ **c** $5r - 3s - 6r + 2s$

2 Simplify these expressions.
  a $y \times y \times y \times y$ **b** $-4r \times 3r$ **c** $-5t^2 \times -3t$

3 Multiply out these brackets.
  a $-4(3 - 4r)$ **b** $3(4 - 3g) + 5(7 - 2g)$

4 Multiply out these brackets.
  a $(2g - h)(g + 3h)$ **c** $(3 + f)(7 - 2f)$
  b $(2r - 1)(2r + 1)$ **d** $(5r + 3s)^2$

5 Factorise:
  a $4d - 12$ **c** $2y^2 + y$
  b $15 - 30r$ **d** $3pq - p^2q$

6 Solve these equations.
  a $3x + 4 = 19$ **c** $5 + 6x = 3x + 35$
  b $\dfrac{x}{5} - 9 = 2$ **d** $3x - 13 = 35$

7 Solve these equations.
  a $3x - 11 = 27 - x$ **b** $2x + 14 = 5x - 10$

8 Solve these equations.
  a $6(x - 3) = 24$ **b** $2(3x + 4) = 38$

9 Make the **red** letter the subject of these formulas.
  a $C = mv - mu$ **b** $e = \dfrac{3f - g}{2}$

10 Make the **red** letter the subject of these formulas.
  a $E = \frac{1}{2}mv^2$ **b** $r = \sqrt{\dfrac{4s}{9}}$

9 Statistics

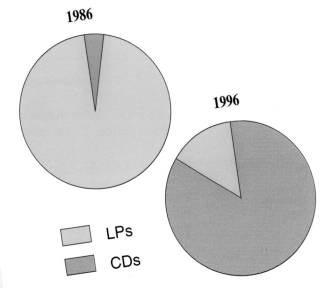

Relative share of retail sales in UK held by LPs and CDs (albums only) in 1986 and 1996.

1986

1996

LPs

CDs

Source: unpublished data.

1 Scatter diagrams

Mammals have different average life spans.
A horse is expected to live about 20 years.
A dog is expected to live about 12 years.
Pregnant mammals carry their unborn young for different lengths of time.
This length of time is called the gestation period.
There is a relationship between the lifespan and the gestation period.

Fiona has found the lifespans and gestation periods for different animals.
She has drawn this graph using her data.

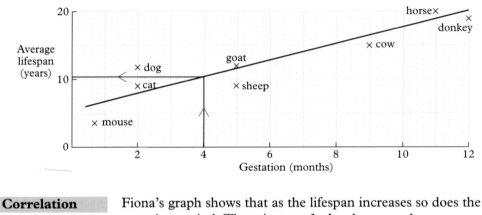

| **Correlation** | Fiona's graph shows that as the lifespan increases so does the gestation period. There is **correlation** between the two. |

| **Scatter graphs** | Graphs like this are called **scatter graphs**. |

| **Line of best fit** | The points seem to lie roughly in a straight line. Fiona has drawn in the line that best fits the points. This line is called the **line of best fit**. |

Fiona knows that the gestation period for a pig is 4 months.
She uses the line of best fit to estimate the average lifespan of a pig.
The red line starts at 4 months. Follow the red line to find an estimate for the average lifespan of a pig.
The estimate is $10\frac{1}{2}$ years.

Exercise 9:1

1 Alan is working on his biology coursework. He is looking to see if there is correlation between the lengths and widths of leaves from a bush. These are his measurements in centimetres.

Length	7.4	6.9	5.0	6.3	6.6	6.5	5.0	6.3
Width	3.5	3.2	2.1	2.8	2.9	3.0	1.8	2.5

a Use graph paper to plot Alan's data.

b Is there correlation? Explain your answer

c Draw the line of best fit.

d Use the line of best fit to estimate the width of a leaf when the length is 6.0 cm

2 Mr Lawson, the maths teacher, has plotted the marks for his maths group in the end of term exams. The pupils sat two papers.

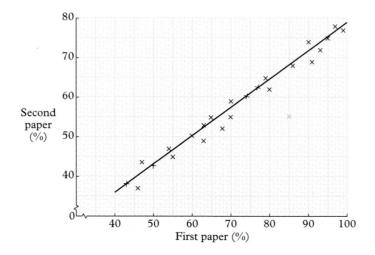

a Joshua sat the first paper but he was ill for the second paper. Estimate what his mark would have been for the second paper if he scored 55% on the first paper.

b Which paper do you think was harder? Explain your answer.

c One of the students was not well when he took the second paper. Use the line of best fit to find the point representing this pupil. What was his mark on the first paper?

3 A doctor is checking the weights of babies at a clinic. He is using this scatter graph to help him identify any babies who are underweight or overweight for their height.

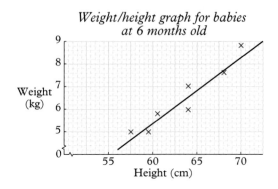

Weight/height graph for babies at 6 months old

Write down for each of these babies whether you think they are underweight, overweight or about the correct weight

Baby	Rashid	Jane	Louise	Sam
Height (cm)	66	63	69	65
Weight (kg)	7.2	5.0	7.9	8.4

There are different types of correlation in these graphs.

Positive correlation

This scatter graph shows the weights and heights of 10 children. As the weight increases so does the height.
This is called **positive correlation**.

Negative correlation

This scatter graph shows sales of gloves and daily temperatures. As the temperature increases so the sales of gloves decrease.
This is called **negative correlation**.

No correlation

This scatter graph shows the salaries and heights of teachers. There is no relationship between the two. The points are scattered all over.
There is **no correlation**.

Correlation can be strong or weak.

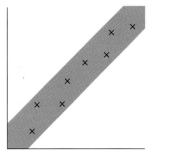

Strong positive correlation
The crosses lie in a narrow band.

Weak negative correlation
The crosses lie in a broad band.

4 Look at each of these scatter graphs.
Describe the type of correlation you can see.

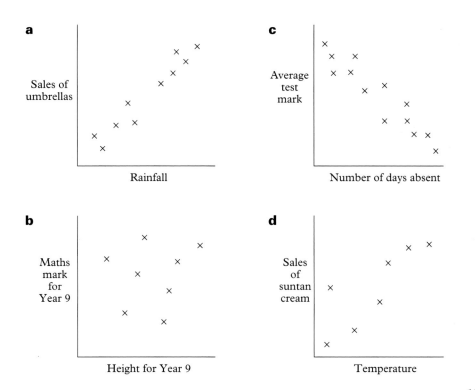

a

Sales of umbrellas

Rainfall

b

Maths mark for Year 9

Height for Year 9

c

Average test mark

Number of days absent

d

Sales of suntan cream

Temperature

5 Laura is doing a survey. She asks eight of her friends how much time they spent on homework last night and how long they watched the television. These are her results. The times are in minutes.

Homework	65	120	85	35	160	100	70	95
Television	190	90	150	210	85	250	150	140

  a Draw a scatter diagram to show Laura's data.
  b Does your graph show any correlation?
     If it does, describe the type of correlation.

6 Jack and Sara manage eight paper shops.
Jack thinks that the weekly sales of papers depend on the size of the shop.
Sara thinks that the sales depend on the number of houses within 3 miles.

They used their sales figures to draw these two scatter graphs.

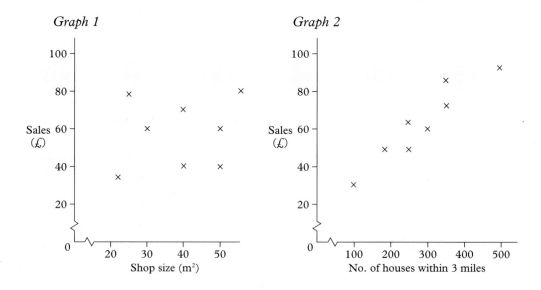

  a What does graph 1 show about the relationship between weekly sales and shop size?
  b What does graph 2 show?
  c Jack and Sara are asked to manage another paper shop.
     It has a floor area of 28 m² and there are 450 houses within 3 miles.
     Use one of the graphs to estimate the sales that the shop is likely to make.
     Write down which graph you used. Explain how you made your estimate.

2 Pie-charts

Below sea level

At sea level

Above sea level

The land in Holland is mostly flat.
The small number of hills are not very high.
The pie-chart shows how the land is divided between below sea level, at sea level, and above sea level.

Exercise 9:2

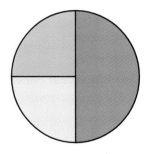

1 The whole circle is 100%.
    a What percentage of the circle is red?
    b What percentage is green?
    c (1) What fraction of the circle is red or yellow?
       (2) What percentage is this?

You may have to *estimate* the percentages shown by 'slices' of pie-charts.

Example   This pie-chart shows what pupils in a school do for lunch.
    a Estimate the percentage that buy lunch in the school canteen.
    b Estimate the percentage that bring a packed lunch.

Packed lunch

School canteen

Go home

    a The 'slice' for school canteen is a bit less than half.
       It is a bit less than 50%.
       An estimate is 45%.
    b The 'slice' for packed lunch is a bit more than a quarter.
       It is a bit more than 25%.
       An estimate is 30%.

2 Estimate the percentage of each of these pie-charts that is coloured red.

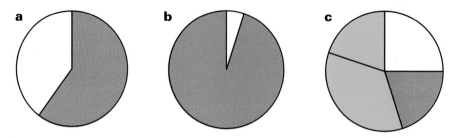

a   **b**   **c**

3 This pie-chart shows the contents of a type of cheese.

Contents	Percentage
fat	31%
water	38%
protein	25%
carbohydrate	

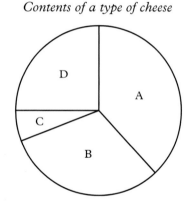

Contents of a type of cheese

a What does part A of the pie-chart represent?
   Use the contents table to help you.
b What does part D represent?
c Use the contents table to find the percentage part C represents.

4 These pie-charts show the percentages of Year 9 girls and boys at Stanthorne High who go home to lunch.

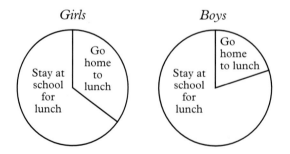

Girls   *Boys*

Girls: Stay at school for lunch / Go home to lunch

Boys: Stay at school for lunch / Go home to lunch

a Estimate the percentage of girls who go home to lunch.
b Estimate the percentage of boys who go home to lunch.
c The number of girls in Year 9 is approximately equal to the number of boys. Sketch a new pie-chart for the whole of Year 9 to show the percentage who go home to lunch.

5 The pie-chart shows the amount of each type of food recommended for a healthy diet.

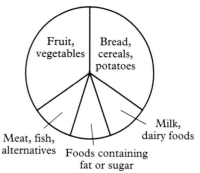

Balance of food for a healthy diet

a Estimate the percentage in each section. Make a table to show these percentages

b A man eats 2500 calories a day. Estimate the number of calories he should get from 'bread, cereals and potatoes'.

c A woman eats 2000 calories a day. Estimate the number of calories she should get from 'meat, fish and alternatives'.

◄◄**REPLAY**►

Example

The table shows the percentage of each type of fish bought by the customers of a fish and chip shop one month.

Type of fish	cod	plaice	scampi	haddock
Percentage of customers	47	15	10	28

Show these results in a pie-chart.

1 100% of the circle is 360°, so 1% is 360° ÷ 100 = 3.6°

2 Work out the angle for each fish. This is easy to do in a table.

Fish	Number of people	Working	Angle
cod	47	47 × 3.6°	169.2°
plaice	15	15 × 3.6°	54°
scampi	10	10 × 3.6°	36°
haddock	28	28 × 3.6°	100.8°
Total	100		360°

3 Check that the angles add up to 360°.

4 Draw a circle and mark the centre.
Draw a line to the top of the circle.
Draw the first angle.
Estimate any parts of degrees.
Carry on until you have drawn all the angles.

Label the finished pie-chart.

Type of fish

183

Exercise 9:3

1 The school canteen service worked out the percentage of each type of vegetarian meal sold during a term.
The results are recorded in the table.

Meal	cheese flan	vegeburger	vegetable curry	bean casserole
Percentage of sales	34	13	24	29

Draw a pie-chart to show the meals sold.

2 A local authority did a survey to find out what secondary school pupils did for lunch. They worked out the percentages of pupils eating in school canteens, bringing a packed lunch and those leaving school to eat elsewhere. They compared their results with those of a similar survey done 15 years earlier.
a Draw a pie-chart for each set of data.
b Write a sentence comparing the lunch habits of pupils in 1996 with those in 1981.

	1996	1981
School canteen	40%	60%
Packed lunch	25%	27%
Eat outside school	35%	13%

3 *Teen Style* magazine asked their readers to name their favourite type of feature. The numbers of readers liking different features are shown in the table.

Type of feature	fiction	fashion	health	sport	travel
Number of readers	743	607	116	340	194

Draw a pie-chart for the data.

4 Rajiv is doing a survey on cars that fail their MOT.
He has collected data on cars that failed for just one reason.

Reason	brakes or lights	tyres	mechanical	exhaust gas
Number of cars	33	21	51	9

Draw a pie-chart for the data.

3 Misleading diagrams

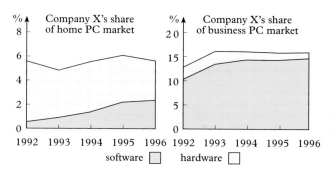

Sometimes charts are used to mislead people.
Changing the scale of a diagram can have a big effect on its appearance. When you read statistical diagrams, you should always look carefully at the scale.

Exercise 9:4

1 Robert sells mountain bikes. He wants to expand his business.
He needs to borrow money from the bank. He wants to show the bank manager that his sales are growing fast. These are the sales for the last 6 months.

Month	Jan	Feb	Mar	Apr	May	June
Number of bikes sold	26	27	29	34	44	55

Robert draws two bar-charts. Chart A uses the whole scale.
In chart B the scale starts at 25.
The two graphs show exactly the same information.
They look very different because the scales are different.

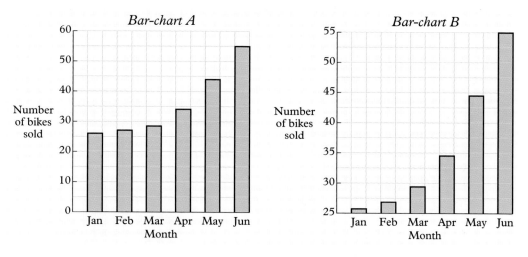

Which chart should Robert show the bank manager?
Explain your answer.

2 This table shows the sales of a company which provides take-away meals. It lists the number of meals sold in the last eight weeks.

Week 1	440	Week 5	465
Week 2	455	Week 6	485
Week 3	450	Week 7	490
Week 4	460	Week 8	495

The manager of the company wants to employ more staff. She wants to show that sales are increasing.

The owner of the company does not want to pay for any more staff. She wants to show that the sales are about the same level.

Draw one graph for each person.
Choose your scale carefully and say who would use each graph.

3 Richard started a part time window cleaning business eight years ago. He now wants to sell the business. He wants to make it look as successful as possible. Here are his profits for the eight years.

Year	1	2	3	4	5	6	7	8
Profit	£200	£225	£235	£270	£290	£330	£370	£335

Draw a bar-chart that would help Richard to make his business look very successful.

4 A computer company is advertising its products. It uses this graph in its advert to show how successful the company is. How is the graph misleading?

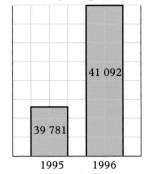

Number of computers sold

41 092

39 781

1995   1996

5 This chart shows a company's profits in 1995 and 1996. Why is the chart misleading?

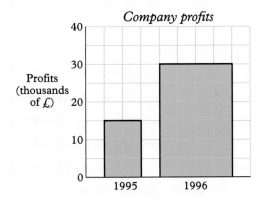

Company profits

6 The Sunshine Holiday Company uses these diagrams to show how its sales are increasing.

Year	1995	1996
Sales	42 000	80 000

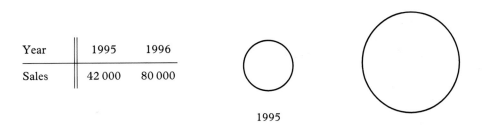

1995

1996

Measure the diameters of these circles. How are these diagrams misleading?

7 A farm selling cheese uses these diagrams to show that sales of cheese have doubled in the last year. What comments can you make about the diagrams?

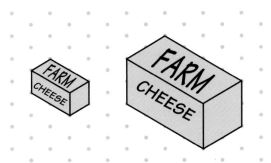

Spending time

Melissa is interested in the way teenagers spend their time.
She guesses that *the more money teenagers get the more time they spend shopping.*
This guess is called a hypothesis.

Investigate the way teenagers spend their time.
Include in your data some information to test Melissa's hypothesis
on shopping.

Make a hypothesis of your own and test it.

Write a report about how teenagers spend their time.
You should include diagrams and calculations.

Remember these rules for questionnaires from last year.

● Questions should not be biased or upset people.

● Questions should be clear and useful to your survey.

● Don't ask questions that give a lot of different answers.

● Put your questions in a sensible order.

1 Steven sells cold drinks in the park.
He recorded the temperature and the number of drinks he sold each day for two weeks. This is his data.

Number of drinks sold	40	50	85	90	100	125	115
Temperature °C	8	9	10	11	11	12	12

Number of drinks sold	150	160	200	255	250	275	290
Temperature °C	13	14	16	18	18	19	20

a Use graph paper to plot Steven's data.
b Describe the correlation.
c Draw the line of best fit.
d Estimate the sales on a day when the temperature is 15°C.
Use the line of best fit.

2 Jean did a traffic survey outside her school.
She recorded how many people were in each car.
These pie-charts show her data for two times of the day.

Between 8.15 and 9.00 am   *Between 9.00 and 9.45 am*

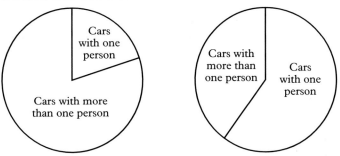

a Estimate what percentage of cars between 8.15 and 9.00 am contained one person.
b Jean surveyed 80 cars between 8.15 and 9.00 am.
About how many contained one person?
c Jean surveyed another 80 cars between 9.00 and 9.45 am.
Estimate the percentage of cars with just one person.
d How did the number of cars with one person alter between the two times?
Can you suggest a reason for this?
e Sketch a new pie-chart showing the percentage of cars containing one person for the total time from 8.15 to 9.45 am.

189

3 Jason carried out a survey of taxi drivers.
He drew these graphs to show his results.

a What does this graph show about the relationship between the numbers of hours worked and the amount of money earned?

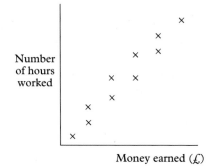

Money earned (£)

b What does this graph show about the relationship between the number of hours worked and the number of hours spent watching television?

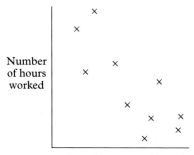

No of hours watching TV

c This table shows Jason's data on the number of hours worked and the number of miles driven. Draw a scatter graph to show this data.

Hours worked	23	29	20	33	35	25
Miles driven	690	826	640	974	980	763

Hours worked	28	32	30	24
Miles driven	850	928	1020	672

Draw the line of best fit on your graph.
Use your line of best fit to estimate the number of miles driven by a driver who works 27 hours.

4 Stacey's parents have told her that if there is a great improvement in her test marks then they will buy her a computer game. These are Stacey's marks in her last five monthly tests.

Jan	Feb	Mar	Apr	May
50%	58%	61%	66%	69%

a Stacey decides to use a graph to show off the increase in her marks. Draw a suitable graph for Stacey. Choose your scale carefully.

b Stacey's rotten brother doesn't want Stacey to have a computer game. He draws a different graph to show Stacey's marks. Draw the graph that Stacey's brother would draw.

5 Sandhills golf club runs a tournament each year. Last year the prize money was £2000. This year it is £4000. The club uses this poster to advertise the tournament. Why is the poster misleading?

6 This graph shows the monthly profits of the school enterprise company. Explain why the graph is misleading.

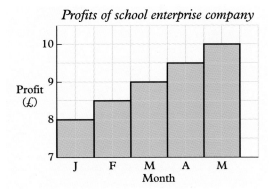

Profits of school enterprise company

7 This table shows how the value of a car changes as it gets older.

Age (years)	1	2	3	5	6	7
Value (£)	9500	7800	7100	4000	3100	1700

a Draw a scatter graph to show this data.
b Describe the correlation.
c Draw the line of best fit.
d Estimate the value of the car when it is 4 years old.
e Estimate in what year the car will have no value. Comment on your answer.

1 The table shows the fuel consumption in miles per gallon of different size car engines.

Fuel consumption (mpg)	Engine (cc)	Fuel consumption (mpg)	Engine (cc)	Fuel consumption (mpg)	Engine (cc)
33	1250	35	1300	32	1400
30	1500	32	1500	29	1600
30	1600	27	1800	28	1800
24	1900	25	1900	22	2000
25	2000	23	2000	20	2200
22	2200	20	2500		

a Use graph paper to draw a scatter diagram of this data.
b Describe the correlation.
c Calculate the mean fuel consumption and engine size.
   Plot this point on your scatter diagram.
d Draw the line of best fit.
   Make it go through the mean point.
e Estimate the fuel consumption of a 1700 cc engine.
   Use the line of best fit.

2 The table shows the number of employees of a local council at the beginning of the year.

Area	Number
teachers and lecturers	6635
education support staff	3386
social services	2532
police and firemen	1447
library	289
others	1857
Total	16 146

a Make a table showing the percentage of employees in each category.
b Draw a pie-chart for the data.
c Find the ratio of education support staff to teachers and lecturers.
   Give your answer in the form $1 : n$.
   Give your answer correct to 3 sf.

- **Correlation**
This graph shows that as the lifespan increases so does the gestation period. There is **correlation** between the two.

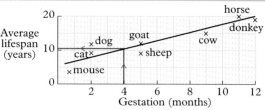

Scatter graphs
Graphs like this are called **scatter graphs**. The points seem to lie roughly in a straight line.

Line of best fit   The line is called the **line of best fit**.

- Correlation can be strong or weak.

Strong positive correlation   Weak negative correlation   No correlation

---

- You can use a pie-chart to show the results of a survey.

Example	Type of fish	cod	plaice	scampi	haddock
	Percentage of customers	47	15	10	28

Show these results in a pie-chart.

1 100% of the pie is 360°, so 1% is 360° ÷ 100 = 3.6°

2 Work out the angle for each fish.
This is easy to do in a table.
e.g. The angle for cod is   47 × 3.6° = 169.2°

3 Check that the angles add up to 360°.

4 Draw and label the pie-chart.

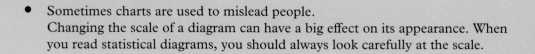

- Sometimes charts are used to mislead people.
Changing the scale of a diagram can have a big effect on its appearance. When you read statistical diagrams, you should always look carefully at the scale.

---

1 The table gives the marks of 12 pupils in their Year 9 examination papers for maths. Both papers are marked out of 50.
  a On graph paper draw a Paper 1 axis across the page and a Paper 2 axis up the page.
  Make both axes go up to 50.

Paper 1	14	36	25	48	39	28	47	40	26	34	18	32
Paper 2	12	37	20	46	40	23	41	35	19	21	19	29

  b Plot the data. Draw the line of best fit.
  c Which paper do you think is easier?
  d Murray got a worse mark than expected for Paper 2.
  What is Murray's Paper 2 mark?
  e Brenda was absent for Paper 1 and got 30 on Paper 2.
  Estimate a Paper 1 mark for Brenda.

2 Penny asked her class where they would like to go on holiday.
She drew this pie-chart to show her results.
  a What fraction of her class chose Australia?
  b About what percentage of her class chose Europe?
  c There are 28 pupils in Penny's class. How many chose America?
  d Pritesh did a similar survey for his class.

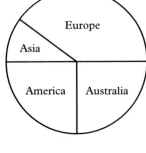
Where my class want to go on holiday

Holiday choice	Europe	Australia	America	Africa	Asia
Percentage of pupils	35	20	24	6	15

Draw a pie-chart to show Pritesh's results.

3 Sandra has drawn this graph to show the amount collected for charity by 9W in the Autumn Term and the Spring Term. Why is the graph misleading?

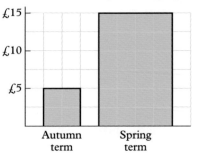
Amount collected for charity by 9W

10 Trigonometry

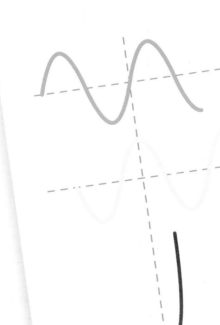

Sine comes from the Latin word *sinus*, meaning a bend or a curve.

Cosine has the prefix *co-*, meaning complement. When the sine curve goes up, the cosine curve goes down.

Tangent comes from the Latin word *tangere* meaning to touch.

1 Introduction

Trigonometry is all about finding lengths and angles in triangles. It is used a lot in building and surveying.

Exercise 10:1

1 Look at this right-angled triangle.
It has a 30° angle at the base.

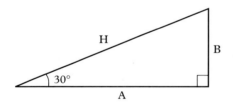

a Draw a right-angled triangle like this.
The angles must be the same as this.
It does not matter how long the sides are.
b Measure all three sides of the triangle to the nearest millimetre.
Be as accurate as you can.
Mark the lengths on your triangle.
c Draw two more triangles that have the same angles as before.
Make the lengths of the sides different.
Mark the lengths on each side. Number your triangles.
d Record your results in a table like this.
Leave the last three columns blank.

Diagram number	Length A	Length B	Length H			

2 **a** For each of the triangles, calculate length **B** ÷ length **A**.
Fill in the answers in column 5 of your table.
Round your answers correct to 3 dp where necessary.

Diagram number	Length A	Length B	Length H	B ÷ A	A ÷ H	B ÷ H

    b What do you notice about these numbers?

    c Make sure your calculator is working in degrees.
Key in: **3** **0** **TAN** . You should get 0.5773502
Write down the answer. Compare it with your answers to B ÷ A.

3 **a** For each triangle, calculate length **A** ÷ length **H**.
Fill in the answers in your table.

    b Fill in the answers to length **B** ÷ length **H** in your table.

    c On your calculator key in: **3** **0** **COS**
Write down your answer.

    d Now calculate **3** **0** **SIN**
Write down your answer.

    e Compare the calculator answers with your table.
Which answer goes with which column?

Tangent The value **3** **0** **TAN** is called the **tangent** of 30°.
It is normally written tan 30° and said 'tan thirty'.

Sine The value **3** **0** **SIN** is called the **sin** of 30°.
It is normally written sin 30° and said 'sine thirty'.

Cosine The value **3** **0** **COS** is called the **cosine** of 30°.
It is normally written cos 30° and said 'cos thirty'.

---

4 **a** Draw a right-angled triangle like this with a 50° angle.

    b Measure the sides and record your results.

    c Work out B ÷ A, A ÷ H and B ÷ H.

    d Compare your answers to **c** with tan 50°, cos 50° and sin 50°.

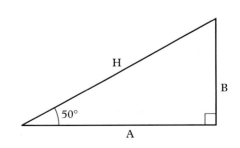

This sort of **trigonometry** only works in right-angled triangles. The values stored in the calculator are very accurate and are much easier than taking measurements.

Sin, cos and tan always have the same value for right-angled triangles with the same angle. The **size** of the triangle does not matter. It is the angles in the triangle that are important.

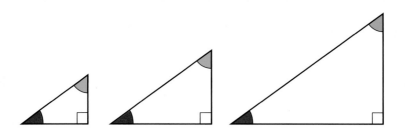

So that it is easy to write the formulas down, the sides of the triangle are given special names.

Hypotenuse   The **hypotenuse** is always the longest side. It never touches the right angle.

Opposite   The **opposite** is the side opposite the angle you are working with. It is one of the two shorter sides.

Adjacent   The **adjacent** is the side next to the angle you are working with. It touches that angle.

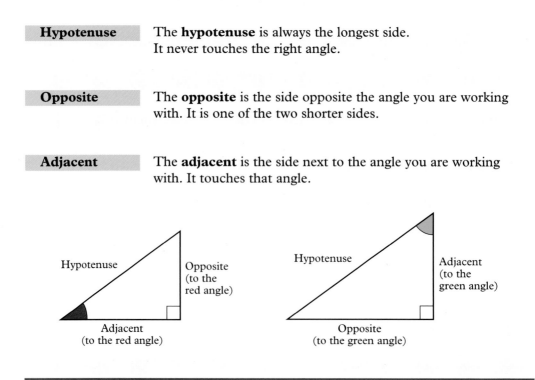

Exercise 10:2

1 Copy this table. Leave space for 10 triangles.

Triangle	Hypotenuse	Opposite to marked angle	Adjacent to marked angle
a	C	D	E
b			

2 For each of these triangles, decide which sides are the **h**ypotenuse, the **o**pposite and the **a**djacent to the marked angle. Fill in the table. The first one is done for you.

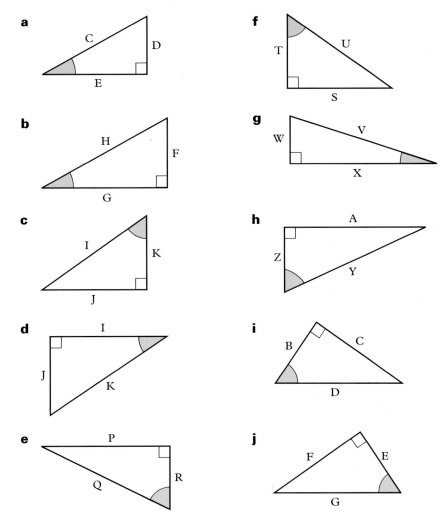

We can now write down the formula for sin, cos and tan using the names of the sides of the triangle.
We will do this for angle a on the diagram.

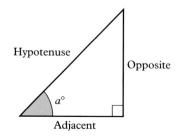

$$\textbf{Sin } a = \frac{\textbf{Opposite}}{\textbf{Hypotenuse}} \qquad \textbf{Cos } a = \frac{\textbf{Adjacent}}{\textbf{Hypotenuse}} \qquad \textbf{Tan } a = \frac{\textbf{Opposite}}{\textbf{Adjacent}}$$

These are often shortened to:

$$\sin a = \frac{\text{opp}}{\text{hyp}} \qquad \cos a = \frac{\text{adj}}{\text{hyp}} \qquad \tan a = \frac{\text{opp}}{\text{adj}}$$

A useful rhyme to remember these is:

The	**Cat**	**Sat**
On	**An**	**Orange**
And	**Howled**	**Horribly**

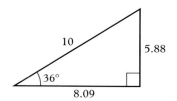

Example   Look at this triangle.
            Work out sin 36°, cos 36° and tan 36°.

$$\sin 36° = \frac{\text{opp}}{\text{hyp}} = \frac{5.88}{10}$$
$$\qquad\quad = \textbf{0.588 to 3 dp}$$

$$\cos 36° = \frac{\text{adj}}{\text{hyp}} = \frac{8.09}{10}$$
$$\qquad\quad = \textbf{0.809 to 3 dp}$$

$$\tan 36° = \frac{\text{opp}}{\text{adj}} = \frac{5.88}{8.09}$$
$$\qquad\quad = \textbf{0.727 to 3 dp}$$

Exercise 10:3

Give all of your answers to 3 dp.

1 Work out sin *a*, cos *a* and tan *a*.

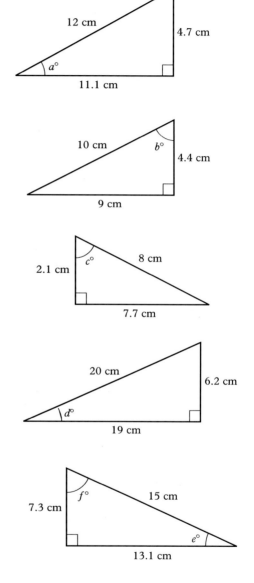

12 cm

4.7 cm

a°

11.1 cm

2 Work out sin *b*, cos *b* and tan *b*.

10 cm

b°

4.4 cm

9 cm

3 Work out sin *c*, cos *c* and tan *c*.

2.1 cm

c°

8 cm

7.7 cm

4 Work out sin *d*, cos *d* and tan *d*.

20 cm

6.2 cm

d°

19 cm

5 **a** Work out sin *e*, cos *e* and tan *e*.
b Work out sin *f*, cos *f* and tan *f*.

f°

15 cm

7.3 cm

e°

13.1 cm

6 Here are the sizes of the angles used in questions **1–5**.
$a = 23°$ $b = 64°$ $c = 75°$ $d = 18°$
$e = 29°$ $f = 61°$.

Use your calculator to check your answers for questions **1–5**.

2 Finding lengths

Jim is painting an upstairs window of a house.
He has a 5 m ladder and wants to know how high it will reach up the wall.
The angle at the base of the ladder is 72°.
He uses trigonometry to work out how far the ladder will reach up the wall.

Example Find the length x in this triangle.

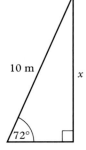

Which length do you know?
The length marked 10 m is the hypotenuse.

Which length do you want to find?
The length marked x is opposite to the 72° angle.

Which formula do you need?

$$\sin a = \frac{\text{opp}}{\text{hyp}} \qquad \cos a = \frac{\text{adj}}{\text{hyp}} \qquad \tan a = \frac{\text{opp}}{\text{adj}}$$

Only the sin formula has opposite **and** hypotenuse in it.

Fill in the formula:
Make sure that your calculator is working in degrees.

$$\sin 72° = \frac{x}{10}$$

$$10 \times \sin 72° = x$$

$$10 \times 0.951 = x$$

$$x = 9.5 \text{ m (1 dp)}$$

You can use the other two formulas in the same way.

Exercise 10:4

Find the lengths marked with letters in each of these triangles.
Round your answers correct to 1 dp. Check that your answers are reasonable.

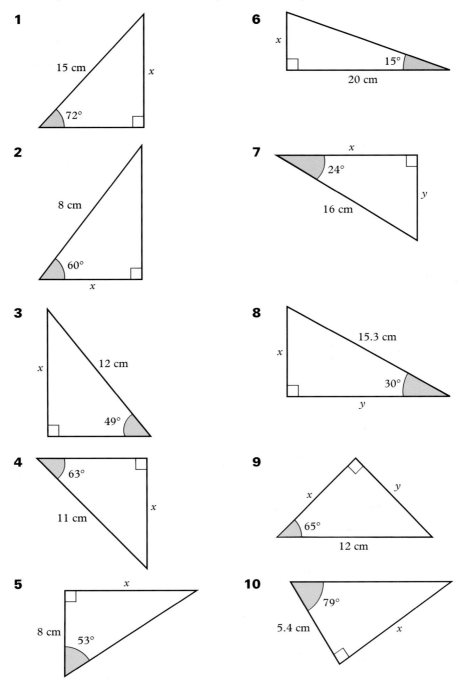

1 15 cm x 72°

6 x 15°   20 cm

2 8 cm   60° x

7 24° x 16 cm y

3 x 12 cm   49°

8 15.3 cm x 30° y

4 63°   11 cm x

9 x y 65°   12 cm

5 x 8 cm   53°

10 79°   5.4 cm x

Exercise 10:5

In this exercise, round your answers correct to 1 dp.

1 Alan's slide is 7 m long. It makes an angle of 36° with the ground. How high above the ground is the top of Alan's slide?

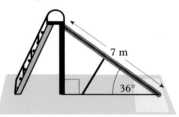

2 The picture shows a roof support. Find the height of the support.

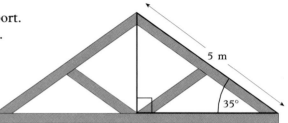

3 The point P is 20 m away from the tree. The angle from the ground to the tree is 24°. Calculate the height of the tree.

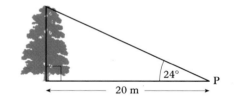

4 a The longest side of a 45° set square is 10 cm. Find the lengths of the other two sides.

   b The longest side of a 30°/60° set square is 15 cm. Find the lengths of the other two sides.

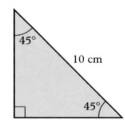
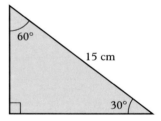

5 A staircase starts 3.5 m away from a wall. It makes an angle of 42° with the wall. How high does the staircase rise?

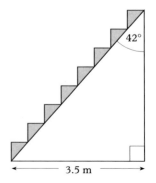

6 A boat is 120 m from the base of the cliff. The angle up to the top of the cliff is 38°.
Work out the height of the cliff.

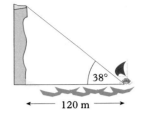

120 m

7 A gardener lays out a triangular plot of land PQR.
The length QR is 15 m, angle P is 90° and angle Q is 30°. Find the lengths of the other two sides of the plot.

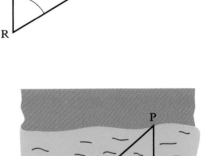

8 The diagram shows three points on the banks of a river.
Use the angle and distance to calculate the width of the river.

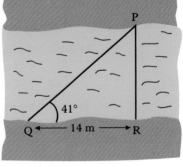

9 Kirsty stands 30 m away from the Telecom Tower in London. She looks up to the top and measures the angle as 80°.
  a Calculate the approximate height of the tower.
    Round your answer to the nearest 10 m.
  b Explain why your answer is inaccurate.

30 m

3 Finding an angle

Trains can only climb very shallow gradients.
You can use trigonometry to find an angle when you know two lengths.
In this case you can find the angle at which the track rises.

If you know the lengths of two of the sides of a right-angled triangle, it is possible to calculate the angles. You use the same formulas.

Example Find the angle marked $a°$ in this triangle.

Which lengths do you know?
The 10 cm side is the hypotenuse.
The 5 cm side is adjacent to $a°$.

Which formula do you need?

$$\sin a = \frac{\text{opp}}{\text{hyp}} \qquad \textbf{cos } a = \frac{\textbf{adj}}{\textbf{hyp}} \qquad \tan a = \frac{\text{opp}}{\text{adj}}$$

Only the cos formula has adjacent **and** hypotenuse in it.

Fill in the formula:

$$\cos a = \frac{5}{10}$$

$$\cos a = 0.5$$

0.5 is the cos of an angle, but which angle?
Make sure that your calculator is working in degrees.
To find out, key in: **0** **.** **5** **SHIFT** **COS**
This should give **60°**

You can use the other two formulas in the same way.

Exercise 10:6

Find the angles marked with letters in each of these triangles.
Round your answers correct to 1 dp.
Check that your answers are reasonable.

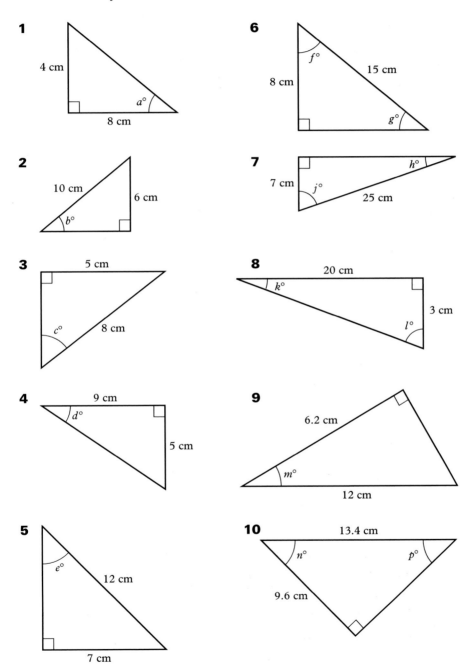

1

4 cm

$a°$

8 cm

2

10 cm

6 cm

$b°$

3

5 cm

$c°$

8 cm

4

9 cm

$d°$

5 cm

5

$e°$

12 cm

7 cm

6

$f°$

8 cm

15 cm

$g°$

7

7 cm

$j°$

$h°$

25 cm

8

20 cm

$k°$

3 cm

$l°$

9

6.2 cm

$m°$

12 cm

10

13.4 cm

$n°$

$p°$

9.6 cm

Exercise 10:7

In this exercise, round your answers to 1 dp.

1 James has a ladder 10 m long.
He wants to reach a point 7 m up
the wall.
What angle should the ladder make
with the ground?

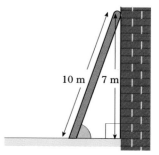

2 A road sign says that this hill has a
gradient of 1 : 4.
This means that it rises 1 m for
every 4 m it goes horizontally.
What angle does the road make with
the horizontal?

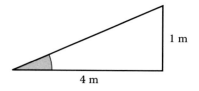

3 A transmitter is supported by wires.
The transmitter is 40 m tall.
The wires are 60 m long.
What angle do the wires make with
the top of the transmitter?

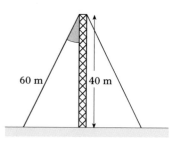

4 A car bonnet has a length
of 1.4 m.
When fully open it is 0.8 m
above the horizontal.
What angle does the car
bonnet turn through?

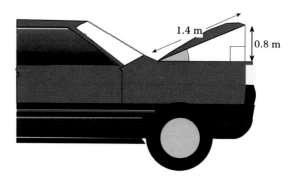

5 Look at the rectangle ABCD.
Find the angle between the side BC
and the diagonal.

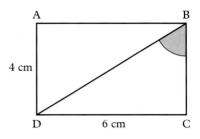

6 This set of step ladders has a height
of 1.6 m.
The ladders are 1.9 m long.
a Find the angle marked $a°$.
b What is the angle between the
ladders?

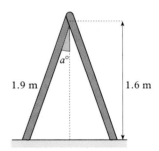

7 Jamie's slide is 8 m long.
The top of the slide is 5 m above the
ground.
What angle does the slide make with
the ground?

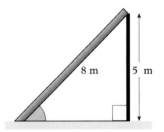

8 Ned is 1.78 m tall.
In the middle of the afternoon his
shadow is 1.32 m long.
What angle do the sun's rays make
with Ned?

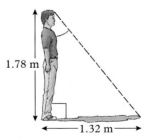

Exercise 10:8

This exercise is a mixture of angle and length problems.
Round all of your answers correct to 1 dp.
In questions **1–4** find the lengths and angles marked with letters.

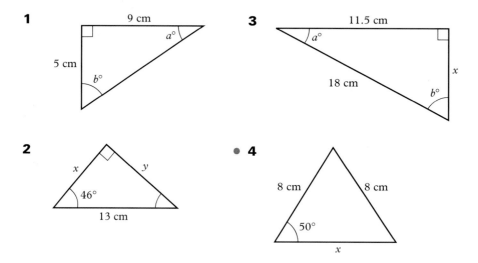

1 9 cm $a°$ 5 cm $b°$

3 11.5 cm $a°$ 18 cm $b°$ x

2 x y 46° 13 cm

• 4 8 cm 8 cm 50° x

5   An aeroplane is 4500 m from touchdown. Its angle of descent is 50° to the horizontal. How high is it above the ground?

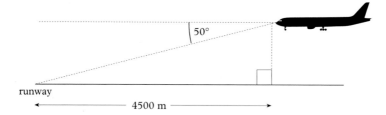

50°

runway

4500 m

6   A ladder is 12 m long. It is put against a wall so that it reaches a height of 10.5 m.
   a What angle does the ladder make with the ground?
   b What angle would allow the ladder to reach 11 m above the ground?
   c The maximum angle the ladder is allowed to make with the ground is 70°. How far can the ladder reach up the wall?

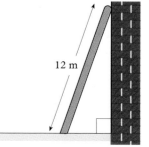

12 m

1 Find the lengths marked with letters in each of these triangles.
Round your answers correct to 1 dp.

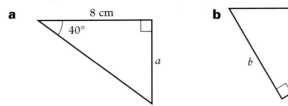

a 8 cm 40° a

b 37° b 23 mm

2 Kamal has drawn a right angled triangle ABC.
  a Write down the value of sin A for Kamal's triangle.
  b Write down the value of cos B.
  c Write down what you notice about your two answers.

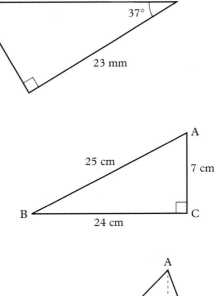

A 25 cm 7 cm B 24 cm C

3 Jason wants to find the area of this triangle.
He starts by working out the height.
  a Find the height.
  b What is the area of the triangle?
    Round your answers correct to 1 dp.

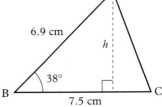

A 6.9 cm h 38° B 7.5 cm C

4 Find the angles marked with letters in each of these triangles.
Round your answers correct to 1 dp.

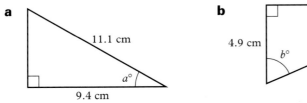

a 11.1 cm $a°$ 9.4 cm

b 4.9 cm $b°$ 8.7 cm

5 Jane is making a bracket for a hanging basket like the one shown.
She works out the angles of the bracket.
Find the angles of the bracket.
Give the angles to a sensible degree of accuracy.

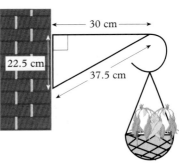

30 cm 22.5 cm 37.5 cm

6 A chord of length 10 cm is drawn in a circle.
The circle has a radius of 6.5 cm.
Calculate the angle that the radius makes
with the chord.

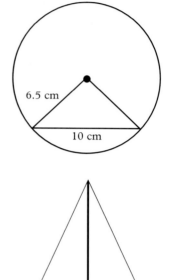

7 A flagpole is held upright by two wires.
Sue estimates that the angle between the
ground and one of the wires is 75°.
She estimates that the wire reaches the
ground 2.5 m from base of the flagpole.
Use Sue's estimates to find the height of the
flagpole. Round your answer to a sensible
degree of accuracy.

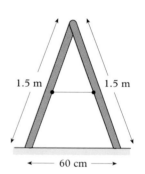

8 The diagram shows a pair of steps.
Each side of the steps is 1.5 m long.
a Find the angle between the sides of
the steps.
The foot of one side is 60 cm from
the foot of the other side.
b The mid-point of each side is joined
by a rope.
Find the length of the rope.

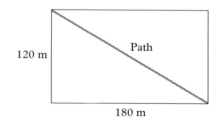

9 A rectangular field is 180 m long and
120 m wide. A path runs diagonally
across the field.
a Calculate the angle that the path
makes with one of the shorter sides
of the field.
Give the angle correct to the nearest
degree.
b Use Pythagoras to calculate the
length of the path.
Give your answer correct to the
nearest metre.

1 Sally and Pritesh are flying a kite.
The kite string is 60 m long.
Sally and Pritesh want to know the height of
the kite.

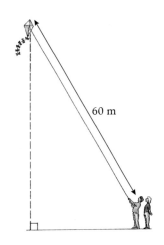

60 m

a Sally estimates that the distance from
Pritesh to the point directly below the
kite is 35 m.
She works out the height using
Pythagoras theorem.
Find Sally's answer.

b Pritesh estimates that the angle the string
makes with the ground is 60°.
He uses trigonometry to work out the height.
Find Pritesh's answer.

c Sally and Pritesh have got different answers.
Write down a sensible estimate for the height of the kite.

2 Use trigonometry to find which of these triangles it would be
impossible to draw. Show your working.

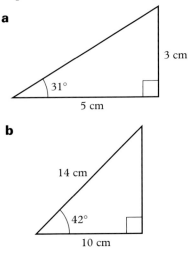

a

31°

5 cm

3 cm

c

58°

7.5 cm

12 cm

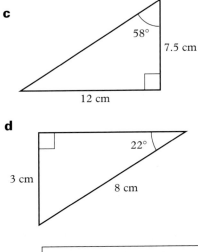

b

14 cm

42°

10 cm

d

3 cm

22°

8 cm

3 A buoy is 7 km east and 4 km north of
a lighthouse.
Use trigonometry to find the bearing of
the buoy from the lighthouse.

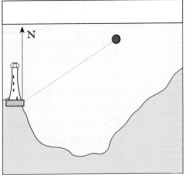

N

4 This rhombus has an angle of 140°
at the top.
Its sides are of length 5 cm.
 a Make a sketch of the rhombus.
 b Draw the two diagonals of the
rhombus on your sketch.
 c Calculate the length of each diagonal.
 d Calculate the area of the rhombus.

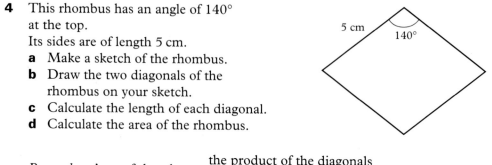

Remember: Area of rhombus = $\dfrac{\text{the product of the diagonals}}{2}$

Give your answers correct to 3 sf.

5 In this question, give your
answers correct to 2 dp.
This right-angled triangle has
angles of 48° and 42°.
Its hypotenuse is 15 cm long.

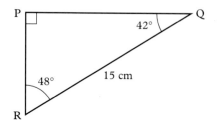

 a Find the lengths PQ and PR.
 b Another right-angled triangle has the same angles.
The length *opposite* its 48° angle is 15 cm.
Draw a diagram to show the second triangle.
 c Find the lengths of the other two sides in the second triangle.

6 The diagram shows the side of a
house.
 a Angle $x°$ is the angle between
the two sloping edges of the
roof.
Calculate angle $x°$ correct to
the nearest degree.
 b Calculate the length of one of
the sloping edges of the roof.
Give the length correct to
1 dp.

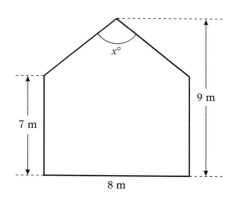

- $\sin a = \dfrac{\text{opp}}{\text{hyp}}$

  $\cos a = \dfrac{\text{adj}}{\text{hyp}}$

  $\tan a = \dfrac{\text{opp}}{\text{adj}}$

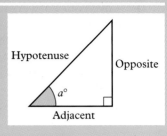

A useful rhyme to remember these is:

The	**Cat**	**Sat**
On	**An**	**Orange**
And	**Howled**	**Horribly**

- *Example*   Find the length x in this triangle.

  Which formula do you need?

  $\sin a = \dfrac{\textbf{opp}}{\textbf{hyp}}$

  $\cos a = \dfrac{\text{adj}}{\text{hyp}}$

  $\tan a = \dfrac{\text{opp}}{\text{adj}}$

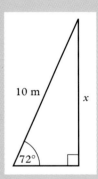

  Only the sin formula has opposite **and** hypotenuse in it.

  $$\sin 72° = \frac{x}{10} \qquad \text{To find } \sin 72°$$

  Key in: **7** **2** **SIN**

  $$0.9511 = \frac{x}{10}$$

  $$0.9511 \times 10 = x \qquad \boldsymbol{x = 9.5\,m\ (1\ dp)}$$

- *Example*   Find the angle marked a in this triangle.

  Which formula do you need?

  $\sin a = \dfrac{\text{opp}}{\text{hyp}} \qquad \cos a = \dfrac{\textbf{adj}}{\textbf{hyp}} \qquad \tan a = \dfrac{\text{opp}}{\text{adj}}$

  Only the cos formula has adjacent **and** hypotenuse in it.

  $$\cos a = \frac{5}{10}$$

  $$\cos a = 0.5$$

  To find $a°$ key in **0** **.** **5** **SHIFT** **COS**   This should give **60°**

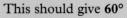

1 Give your answers to 2 dp when necessary.

a For this triangle write down the value of:

    (1) tan A   (2) sin A   (3) cos A

b Write down the value of:

    (1) tan B   (2) sin B   (3) cos B

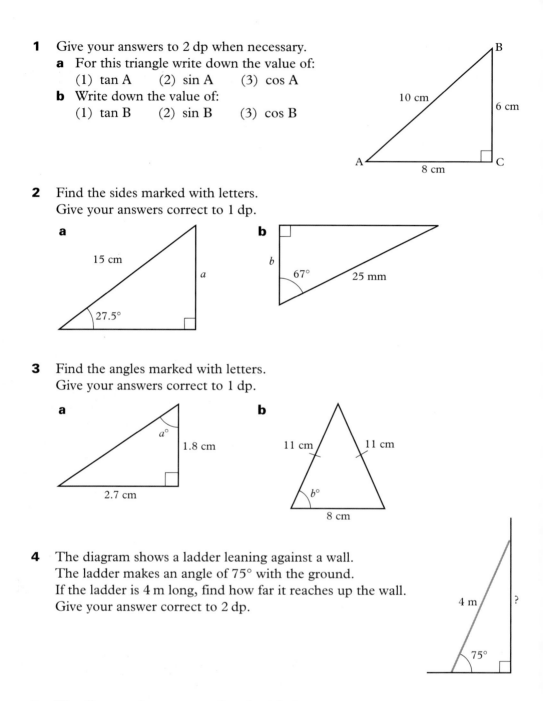

2 Find the sides marked with letters.
Give your answers correct to 1 dp.

a

15 cm

a

27.5°

b

b

67°

25 mm

3 Find the angles marked with letters.
Give your answers correct to 1 dp.

a

a°

1.8 cm

2.7 cm

b

11 cm   11 cm

b°

8 cm

4 The diagram shows a ladder leaning against a wall.
The ladder makes an angle of 75° with the ground.
If the ladder is 4 m long, find how far it reaches up the wall.
Give your answer correct to 2 dp.

4 m   ?

75°

5 The diagram shows a ramp for wheelchairs.
Find the angle that the ramp makes with the ground.
Give your answer correct to the nearest degree.

20 cm

?°

115 cm

11 The best chapter, probably

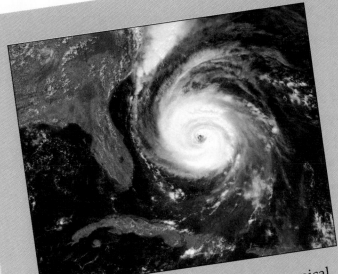

A hurricane is an intense, devastating tropical storm caused by a low-pressure weather system. Hurricanes occur in tropical regions usually between July and October.

Weather forecasters try to warn people of the approach of a hurricane. The forecasters use probabilities to give an idea of the likelihood of a hurricane striking a particular area.

1 Probability and relative frequency

Rudi has bought a raffle ticket.
The prize is a holiday.
The supermarket has sold
thousands of tickets.
Rudi knows his chance of winning
the holiday is very small.

◀◀**REPLAY**▶

Probability	**Probability** tells us how likely something is to happen. All probabilities must be between 0 and 1.

Exercise 11:1

1 A box contains 3 red, 5 green and 4 blue pencils.
  Pat chooses a pencil at random. What is the probability that it is:
  a red
  b green
  c either green or blue
  d neither red nor green?

2 The table shows the membership
  of a youth club.
  a How many members of the club
    are there altogether?
  b A member is chosen at random.
    What is the probability that the
    member is:
    (1) female
    (2) under 16
    (3) male and over 16?

	Under 16	16 and over
Male	18	26
Female	21	25

3 Katy has done a survey on the favourite sports of her friends.
She has drawn a bar-chart to show her data.

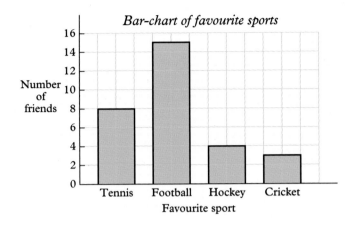

One of Katy's friends is chosen at random.
a What is the probability that the friend's favourite sport is:
   (1) tennis   (2) hockey or cricket?
b Copy this probability scale.

0   1
|——————————————————————————————————|

Mark the probability that the friend's favourite sport is football on
your scale.

4 Michael has 7 white beads and 1 black bead in a bag.
He picks out a bead without looking.

His father says the probability of the bead
being black is $\frac{1}{7}$ because there are 7 white
beads and one black bead.
His mother says the probability is $\frac{1}{8}$ because
there are 8 beads and just one is black.

a Whose answer is correct?
   Why is the other answer wrong?
b Michael changes the number of beads in the bag.
   The probability of choosing a black bead is now $\frac{3}{11}$
   How many black beads and how many white beads could be in
   the bag now?
   Is this the only possible answer? Give your reasons.

Probabilities always add up to 1.

Example

The probability that Peter will go bowling sometime during this week is $\frac{3}{5}$.

What is the probability that Peter will not go bowling?

Probability that Peter will not go bowling $= 1 - \dfrac{3}{5}$

$$= \dfrac{2}{5}$$

Exercise 11:2

1 The probability that the Khan family will go to France for their holiday is $\frac{8}{9}$. What is the probability that they will not go to France for their holiday?

2 Out of every 100 people, 13 are left handed.
What is the probability that a person chosen at random is:
a left handed　　　　　　　　**b** not left handed?

3 Pauline is using this spinner in a game.
What is the probability that it will land on:
a red　　　　　**c** either red or green
b blue　　　　　**d** a colour other than green?

4 Liam has to decide what sport to choose for his activity afternoon.
The probability of each of his choices is given in the table.

Football	Climbing	Swimming
0.40	0.25	0.35

a Write down the probability that he chooses:
　(1) football
　(2) climbing or swimming
b What is the probability that he does not choose swimming?

Event	The word **event** is often used in probability. An event is one thing that can happen in an experiment.
	When you choose a card at random from a pack of 52 playing cards there are lots of possible events. Some examples are: getting a 9 of hearts, getting a red card.
	We can call an event by a letter. The event 'getting a heart' might be called event A. We write $P(A)$ for the probability of A happening.
	The probability of getting a heart is $\frac{13}{52}$.
	So we write $P(A) = \dfrac{13}{52} = \dfrac{1}{4}$
Probability of something not happening	The **probability of something not happening** is 1 minus the probability that it does happen. For an event A, the event 'not A' is written A'.
	$P(A') = 1 - P(A)$
	The probability of not getting a heart, $P(A') = 1 - \dfrac{1}{4} = \dfrac{3}{4}$

Exercise 11:3

1 Find $P(A)$ if the event A is getting:
   a a red ball from a bag containing 3 red balls and 5 blue balls
   b a number more than 4 on a fair dice
   c an ace from a pack of 52 playing cards.

2 Find $P(A')$ for each of the events in question **1**.

3 Mark says that the probability that his team will win their match is 0.6
   The probability that they will lose is 0.3
   Let the event A be 'the team will draw the match'.
   a Find $P(A)$ **b** Find $P(A')$

4 The probability that Samina goes home from the disco by bus is 0.35, by taxi is 0.45
   Let A be the event that Samina 'goes home by another method'.
   Find   **a** $P(A)$ **b** $P(A')$

Challenge

This is for 2 or more people.

You need a bag containing 10 beads.
Do not look inside the bag.

Choose a bead from the bag without looking.

Note the colour of the bead on a
copy of this tally-table. Replace
the bead. A second person has a
go, notes the colour and replaces
the bead. This is done 10 times.

Colour	Tally	Total

Use the data collected to guess the contents of the bag.

Now repeat the sampling another 10 times.
Use both sets of data to improve your guess.

Repeat the experiment several times until you think that you have
worked out the correct contents of the bag.
Is your final answer more reliable than your first answer?
Give a reason why.

Check your answer by looking at the contents of the bag.

Ask your teacher for another bag.
Find the contents of the bag using the method above.

......

Jane has thrown her dice 100 times.
She has recorded the number each time using tally marks.

Number	Tally	Frequency
1	ЖЖ Ж IIII	19
2	Ж Ж Ж	15
3	Ж Ж Ж I	16
4	Ж Ж Ж III	18
5	Ж Ж Ж II	17
6	Ж Ж Ж	15

Jane has labelled the total column 'Frequency'.

| **Frequency** | The **frequency** of an event is the number of times that it happens. |

Jane thinks that the dice is fair.

If it is fair, the probability of getting each number is $\frac{1}{6}$

She checks her data to see if the dice is fair.
She uses the relative frequency from her experiment.

| **Relative frequency** | The **relative frequency** of an event $= \dfrac{\text{frequency of the event}}{\text{total frequency}}$ |
| | The relative frequency gives an *estimate* of the probability. |

Jane's data shows that the frequency of getting 1 on her dice is 19.
The total of all the frequencies is 100.

The relative frequency of $1 = \dfrac{19}{100}$

Jane wants to see if $\frac{1}{6}$ and $\frac{19}{100}$ are approximately the same value.

She converts them both to decimals.

$$\frac{1}{6} = 0.17 \text{ (2 dp)} \qquad \frac{19}{100} = 0.19$$

The two values are very close.

Jane can get a better estimate for the probability by repeating the experiment more times.

Exercise 11:4

1 Ceri tossed two coins 100 times.
This is her data.

Outcome	Tally	Frequency
2 heads	ЖЖ ЖЖ ЖЖ ЖЖ ‖	22
2 tails	ЖЖ ЖЖ ЖЖ ЖЖ ЖЖ ‖	27
head, tail	ЖЖ ЖЖ ЖЖ ЖЖ ЖЖ	
	ЖЖ ЖЖ ЖЖ ЖЖ ЖЖ ǀ	51

Find the relative frequency of
a 2 heads **b** 2 tails **c** a head and a tail
Give your answers as decimals.

2 Simon collected data on the colours of cars passing the school gate.
His results are shown in the table.

Colour	Tally	Frequency
black	ЖЖ ‖‖‖	9
red	ЖЖ ЖЖ ЖЖ ЖЖ ЖЖ ЖЖ	30
white	ЖЖ ЖЖ ЖЖ ЖЖ ЖЖ ǀ	26
blue	ЖЖ ЖЖ ‖	12
green	ЖЖ ЖЖ ЖЖ ‖‖‖	19
other	‖‖‖	4

a How many cars did Simon include in his survey?
b What is the relative frequency of white?
   Give your answer as (1) a fraction (2) a decimal (3) a percentage
c What is the relative frequency of black?
   Give your answer as (1) a fraction (2) a decimal (3) a percentage
d What is the most likely colour of the next car passing the school gate?
e Write down an estimate for the probability that the next car will be green. Give your answer as a fraction.
f How can the estimate for the probability of green be made more reliable?

3 Jennie has carried out an experiment to see which way a drawing pin lands when it is dropped on to the floor. Here is her data.

Position	Tally	Frequency
pin up	৷৷৷ ৷৷৷ ৷৷৷ ৷৷৷ ৷৷৷ ৷৷৷ ৷৷৷ ৷	36
pin down	৷৷৷ ৷৷৷ ৷৷৷ ৷৷৷ ৷৷৷৷	24

    a How many times has Jennie dropped the pin?

    b Work out the relative frequencies of each way the pin lands.
       Give your answers as decimals.

    c Which way is the pin more likely to land?

    d How could Jennie be more sure of her prediction?

4 Jason is designing a game for his school charity day.
Pupils pay to take part.
They roll a penny on to this board and win prizes depending on where the penny ends up.
If the penny lands completely inside a red square they win the amount of money written in that square.

Jason needs to know how often pupils will win.
He can then decide on an entry fee and the value of the prizes.
He decides to roll pennies and record the outcomes.
Here is his data.

Outcome	Tally	Frequency
win	৷৷৷ ৷৷৷ ৷৷	12
lose	৷৷৷ ৷৷৷ ৷৷৷ ৷৷৷ ৷৷৷ ৷৷৷ ৷৷৷ ৷৷৷ ৷৷৷	38

    a Give an estimate for the chance that a pupil will win with one go.

    b Jason decides to charge pupils 5 p to take part in his game.
       How much should he give as a prize?
       Remember that he is trying to raise money for charity.

2 Listing outcomes

This stall is a popular money raiser. The person running the stall takes out insurance against someone actually winning the car. The insurance company uses probability to work out the chances of throwing six 6s in a row. They have to provide the car if someone wins!

◄◄REPLAY►

Sample space	A **sample space** is a list of all the possible outcomes.
Sample space diagram	A table which shows all of the possible outcomes is called a **sample space diagram**.

Example John tosses this coin and spins this spinner.
a Draw a sample space diagram.
b Write down the probability that John gets a tail and a 3.

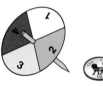

a

		Number on spinner			
		1	2	3	4
Coin	H	H, 1	H, 2	H, 3	H, 4
	T	T, 1	T, 2	T, 3	T, 4

b There are eight possible outcomes shown in the diagram. T, 3 appears once. The probability of a tail and a 3 is $\frac{1}{8}$.

Exercise 11:5

1 Copy this sample space diagram. Fill it in to show all the possible outcomes of tossing two coins.

	50 p coin	
	H	T
20 p coin H		
T		

2 A coin is tossed and a dice thrown.
Copy and complete the sample
space diagram to show all the
possible outcomes.

		Dice					
		1	2	3	4	5	6
Coin	H						
	T						

a What is the total number of possible outcomes?
b What is the probability of getting a head and a three?
c What is the probability of getting a tail and an even number?

3 Marie has to choose one boy and one girl from these pupils:
Rajiv Jim David Karen Sapna Jane
List all her possible choices.

4 Kate, Pam and Wendy take part in a quiz. To decide the order in
which they answer questions their names are put into a bag.
The names are taken out of the bag one at a time.
a List all the ways the names could be drawn out.
b Write down the probability that Wendy is last.

5 Two dice are thrown.
Draw a sample space diagram to
show all the possible outcomes.
a What is the probability of
getting two sixes?
b What is the probability of
getting a 6 and a 4?
c What is the probability of
getting two even numbers?
d What is the probability of getting two numbers that add up to 8?
e What is the probability of throwing a six on the first dice?
What is the probability of throwing a six on the second dice?
Multiply the two probabilities.
Look at your answer to part **a**. What do you notice?

6   Alan and Fatima are doing a probability experiment.
Alan picks a bead at random from his bag.
His bag contains 2 red and 4 blue beads.
Fatima picks a bead at random from her bag.
Her bag contains 1 red and 2 blue beads.
a   Copy and complete this sample space diagram.

		Alan					
		R	R	B	B	B	B
Fatima	R						
	B						
	B						

b   What is the probability of getting two red beads?
c   What is the probability of getting two blue beads?
d   What is the probability that Alan will pick a red bead?
What is the probability that Fatima will pick a red bead?
Multiply these two probabilities.
Both of these events must happen to get the event 'two red beads'.
Look at your answer to part **b**. What do you notice?
e   Now do the same for the two blue beads. What do you notice?

Independent events	When you throw two dice, the number that comes up on one dice has no effect on the number that comes up on the second dice. These are two **independent events**.

Exercise 11:6

Look at the events in questions **1–8**.
Decide whether each pair of events is independent.

1   'A head on tossing a coin' and 'a six on throwing a dice'.

2   Nathan picks a sweet at random from a bag and replaces it.
Ellen then picks another sweet at random from the bag.

3   Nathan picks a sweet at random from a bag and eats it.
Ellen then picks a sweet at random from the bag.

4 Cara looks at the weather out of the window when she gets up.
Cara decides what clothes to wear.

5 Paul decides what to wear to school.
His mother decides what to wear to work.

6 Daniel decides what to eat for tea.
He then decides which homework to do first.

7 Petra decides whether to walk to school or take the bus.
She then decides what time to leave for school.

8 **a** Write down two events that are independent.
b Write down two events that are not independent.

Probability of independent events

If two **events** A and B are **independent** then the **probability** of them *both* happening is called $P(A \text{ and } B)$

$$P(A \text{ and } B) = P(A) \times P(B)$$

Example

a Asha spins this spinner once.
What is the probability of getting blue?

b Asha spins the spinner twice.
(1) What is the probability of getting blue both times?
(2) What is the probability of getting blue the first time and red the second time?

a $P(\text{blue}) = \dfrac{3}{8}$

b (1) $P(\text{blue and blue}) = P(\text{blue}) \times P(\text{blue})$

$$= \frac{3}{8} \times \frac{3}{8} \qquad\qquad = \frac{9}{64}$$

(2) $P(\text{blue and red}) = P(\text{blue}) \times P(\text{red})$

$$= \frac{3}{8} \times \frac{2}{8} \qquad\qquad = \frac{6}{64}$$

Exercise 11:7

1 Two coins are tossed. Find the probability of getting:
    a both coins showing heads
    b a head on the first coin and a tail on the second.

2 The probability that the village football team will win a match is $\frac{2}{3}$
    Find the probability that:
    a the team will win the next two matches
    b the team will lose the next match but win the following one.

3 Robert picks a counter at random from a bag and then replaces it.
    He does this twice.
    The bag contains 5 red, 7 green and 8 black counters.
    Find the probability that he picks out:
    a two red counters
    b a red followed by a black.

4 The probability that Ben remembers his father's birthday is 0.2
    What is the probability that:
    a Ben remembers two birthdays running
    b Ben remembers the next birthday but forgets the following one?

5 Susan has carried out an experiment to find the probability that a
    dropped drawing pin will fall point down.
    Her estimate of the probability is 0.2
    Susan drops three drawing pins.
    Use Susan's estimate to find the probability that:
    a all three drawing pins will land point down
    b all three will land point up
    c only one will land point down.

6 The youth club is running a fund raising stall.
    If you can throw 6 sixes in a row with a dice you win a car.
    Write down the probability of throwing:
    a 2 sixes in a row
    b 3 sixes in a row
    c 6 sixes in a row.

3   Tree diagrams

John is climbing this tree.
When the branches divide, he
can choose which way to go.
There are lots of different ways
of getting to the top.

You can use the idea of a branching tree
in probability to show the different things
that can happen. You will usually draw
the tree on its side like this, with the
branches going from left to right.

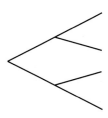

Tree diagrams   You can use **tree diagrams** to show the outcomes of
independent events.

Example   Peter goes to Archery club.
Peter's probability of hitting the target is $\frac{1}{3}$

He fires two shots.
a   Show all the possible outcomes on a tree diagram.
b   Use the diagram to find the probability of Peter having one
hit and one miss in any order.

a   Show Peter's first shot by drawing the first set of branches of the tree.
Write each probability on its branch.

First shot

$\frac{1}{3}$ ⟋ **HIT**

$\frac{2}{3}$ ⟍ **MISS**

Now add Peter's second shot as a second set of branches.

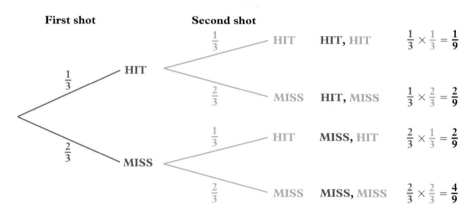

First shot

Second shot

$\frac{1}{3}$ HIT   HIT, HIT $\frac{1}{3} \times \frac{1}{3} = \frac{1}{9}$

$\frac{1}{3}$ HIT

$\frac{2}{3}$ MISS   HIT, MISS $\frac{1}{3} \times \frac{2}{3} = \frac{2}{9}$

$\frac{1}{3}$ HIT   MISS, HIT $\frac{2}{3} \times \frac{1}{3} = \frac{2}{9}$

$\frac{2}{3}$ MISS

$\frac{2}{3}$ MISS   MISS, MISS $\frac{2}{3} \times \frac{2}{3} = \frac{4}{9}$

b Each path through the tree represents a single outcome.
This tree has four paths so there are four possible outcomes.
Two paths give one hit and one miss. Add the outcomes of these two paths.

The probability of Peter having a hit and a miss $= \frac{2}{9} + \frac{2}{9} = \frac{4}{9}$

Exercise 11:8

1 The probability of a pupil buying chips for lunch is 0.7
   a Copy and complete the tree diagram to show the choices of the next two pupils in the queue.

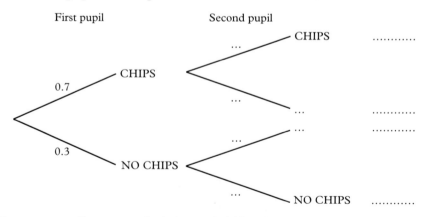

First pupil

Second pupil

... CHIPS ....

CHIPS

0.7

... .

...

...

0.3

NO CHIPS

...

... NO CHIPS ....

Use your tree diagram to find the probability that:
   b each of the next two pupils in the queue will buy chips
   c neither of the next two pupils will buy chips.

2 A garden centre sells mixed crocus bulbs.
The probability of choosing a purple crocus is $\frac{2}{3}$
Copy and complete the tree diagram to show the possible outcomes.

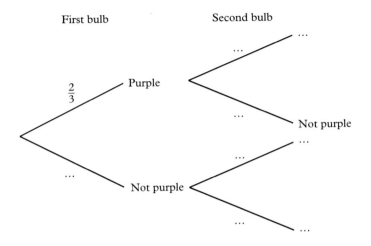

First bulb   Second bulb

Kiran picks two bulbs at random.
Use your tree diagram to find the probability that:
a Kiran's two bulbs are both purple
b only one bulb is purple
c neither bulb is purple.

3 A bag contains 3 white counters and 2 black counters.
Ria picks a counter at random and then replaces it.
She then picks a second counter.
Use a tree diagram to find the probability that:
a both counters are black,
b the first counter is black and the second counter is white,
c only one counter is black.

4 A firm makes computer chips.

The probability of a chip being faulty is $\frac{1}{100}$

Sheila picks two chips at random.
Use a tree diagram to find the probability that:
a both chips are faulty
b only one chip is faulty.

5 Amy comes to school by car. She has calculated that the probability of each set of traffic lights being red is $\frac{2}{5}$
Use a tree diagram to find the probability that:
a both of the next two sets of traffic lights will be red
b neither of the next two sets of traffic lights will be red.

6 Tim and Boris play tennis together.
The probability that Tim will win is 0.6
Use a tree diagram to find the probability that Tim will win:
a both of the next two matches
b only one of the next two matches.

7 The probability that the school football team will win their next match is $\frac{1}{2}$
The probability that they will lose is $\frac{1}{3}$ and that they will draw is $\frac{1}{6}$

a Copy and complete this tree diagram.

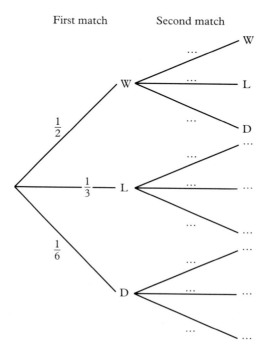

First match   Second match

b Use your tree diagram to find the probability that the team will:
(1) draw both of the next two matches
(2) win only one of the next two matches.

1 Katie is running in a 100 metre race.
There are six runners.

Explain why the probability that Katie will win is **not** $\frac{1}{6}$

2 Make three copies of this spinner.
  a Shade the copies so that each
    shaded portion has the
    probability given.
    (1) Shaded has double the
      chance of unshaded.
    (2) Probability of shaded is 75%.
    (3) Probability of shaded is
      about 40%.
  b Copy this probability scale.

  0   1
  ├──┤

  Show the probability of shaded for each of your three spinners on
  your scale.

3 **a** The probability of getting a 6 on a biased dice is $\frac{1}{4}$

   Write down the probability of not getting a 6.
  b The probability of at least one person in 9M forgetting their maths
    books is 90%
    What is the probability of all of 9M remembering their maths book?

4 Denis has done a survey on the type
  of vehicles passing the school.
  His results are given in the table.

Vehicle	Frequency
car	64
lorry	12
bus	15
motorcycle	9

  a How many vehicles are there altogether in the survey?
  b Write down the relative frequency of each type of vehicle as:
    (1) a fraction   (2) a decimal   (3) a percentage
  c Denis can hear another vehicle coming. What is it most likely to be?

5 The probability that a train arrives early or on time is $\frac{4}{5}$

  a What is the probability that a train is late?
  b What is the probability that the next two trains will be late?

6 Mandy has 20 beads of three different colours in a bag.
She wants to find out how many beads there are of each colour
without looking.
Mandy takes out a bead and writes down the colour.
Mandy then puts the bead back in the bag.
She repeats this 60 times.

The table shows Mandy's results.

Colour	Frequency
red	29
white	11
yellow	20

a Write down the relative frequency of each colour as a fraction.
b How many beads of each colour do you think Mandy has?
c How could Mandy improve her chance of being right about the
numbers of beads?

7 This drinks machine is broken.
You cannot choose which drink you get!

Jane likes all the drinks.
Paul only likes chicken soup.

Jane and Paul buy one drink each.

a Write down the probability that:
(1) Jane will get a drink she likes.
(2) Paul will get a drink he likes.

b Copy this probability scale.

```
0                                                    1
├────────────────────────────────────────────────┤
```

Mark the probabilities from part **a** on your scale.

c Anne buys a drink. The arrow shows the probability that Ann will
get a drink that she likes.

```
                        Ann
0                        ↓                           1
├────────────────────────────────────────────────┤
```

(1) How many of the drinks does Ann like?
(2) Draw a new probability scale.
Mark the probability that Ann will get a drink that she does
not like.

8 Carol puts 12 counters in a bag.
She takes out one counter at random and records the colour.
Carol then replaces the counter.
She does this 12 times.
Here is Carol's tally-table:

Colour	Tally	Total
green	JHT I	6
yellow	IIII	4
red	II	2

  a Carol says, 'There must be 6 green counters in my bag because
         there are 6 greens in my table'.
  Explain why Carol is wrong.
  b What is the smallest number of red counters that can be in the bag?
  c Carol says, 'There cannot be any blue counters in my bag because
         there are no blues in my table'.
  Explain why Carol is wrong.

9 Pam shuffles a pack of cards. She chooses a card, notes the suit and
replaces the card. Pam then offers the pack to Saleem who also
chooses a card.
  a Copy this sample
  space diagram.
  Fill it in.

		Pam's card			
		C	D	H	S
Saleem's card	S				
	H				
	D				
	C				

  b Write down the probability that the cards are:
  (1) a heart and a club   (3) both the same colour
  (2) both red cards   (4) both from the same suit

10 **a** Use the probability tree to
  write down the probability
  that the first person you meet
  will have their birthday in the
  summer.
  b Complete this tree diagram
  for two people.
  c What is the probability that
  the next two people you meet
  will both have their birthdays
  in the summer?
  d What is the probability that
  one will have their birthday
  in the summer but not the
  other?

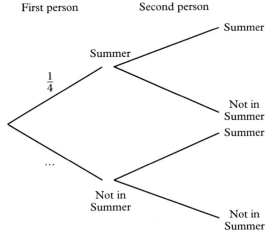

1 A bag contains 4 lemon sweets and 6 orange sweets.

  a David takes a sweet out of the bag without looking.
What is the probability that David chooses a lemon sweet?

  b David eats the sweet and passes the bag to Barbara.
Barbara also takes a sweet without looking and eats it.
Complete the tree diagram to show David and Barbara choosing sweets.

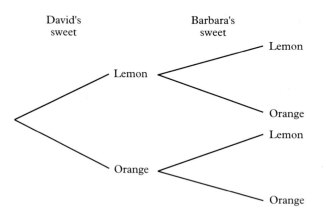

David's sweet   Barbara's sweet

Lemon
 — Lemon
 — Orange

Orange
 — Lemon
 — Orange

  c Use the tree diagram to find the probability that David and Barbara both eat sweets of the same flavour.

2 Sally has these tins of soup which have lost their labels.
She knows that she bought 3 tins of tomato soup and 2 of chicken soup.

Sally opens two of the tins.
Use a tree diagram to find the probability that:

  a both the tins Sally opens contain chicken soup

  b Sally opens one tin of each flavour.

- **Probability** **Probability** tells us how likely something is to happen.

- **Relative frequency** The **relative frequency** of an event $= \dfrac{\text{frequency of the event}}{\text{total frequency}}$

  The relative frequency gives an *estimate* of the probability.

- **Sample space diagram** A table which shows all of the possible outcomes is called a **sample space diagram**.

  Example John tosses this coin and spins this spinner.
  a Draw a sample space diagram.
  b Write down the probability that John gets a tail and a 3.

a

		Number on spinner			
		1	2	3	4
Coin	H	H, 1	H, 2	H, 3	H, 4
	T	T, 1	T, 2	T, 3	T, 4

  b There are eight possible outcomes shown in the diagrams. T, 3 appears once. The probability of a tail and a 3 is $\frac{1}{8}$.

- **Tree diagrams** You can use **tree diagrams** to show the outcomes of independent events.

  Example Peter goes to Archery club. Peter's probability of hitting the target is $\frac{1}{3}$. He fires two shots. Show all the possible outcomes on a tree diagram. Use the diagram to find the probability of Peter having one hit and one miss in any order.

First shot	Second shot		
	$\frac{1}{3}$ HIT	HIT, HIT	$\frac{1}{3} \times \frac{1}{3} = \frac{1}{9}$
$\frac{1}{3}$ HIT	$\frac{2}{3}$ MISS	HIT, MISS	$\frac{1}{3} \times \frac{2}{3} = \frac{2}{9}$
$\frac{2}{3}$ MISS	$\frac{1}{3}$ HIT	MISS, HIT	$\frac{2}{3} \times \frac{1}{3} = \frac{2}{9}$
	$\frac{2}{3}$ MISS	MISS, MISS	$\frac{2}{3} \times \frac{2}{3} = \frac{4}{9}$

  Each path through the tree represents a single outcome.
  This tree has four paths so there are four possible outcomes.
  Two paths give one hit and one miss.
  Add the outcomes of these two paths.
  The probability of Peter having a hit and a miss $= \frac{2}{9} + \frac{2}{9} = \frac{4}{9}$

1 Decide whether each of these statements is true or false:
   a Year 9 can choose to study French or German.
       There are 2 choices so the probability that a pupil chooses French is $\frac{1}{2}$.
   b An ordinary dice is thrown. There are 6 numbers so the probability
       of getting 5 is $\frac{1}{6}$.

2 The probability that a Year 9 pupil chosen at random has a scientific
   calculator is 0.8. Write down the probability that the same Year 9
   pupil does not have a scientific calculator.

3 A take-away has noted the first 100 meals bought one Saturday evening.
   Here are their results:

Meal	Frequency
chicken and chips	35
fish and chips	31
chicken curry	14
prawn fried rice	20

   a Write down the relative frequency of fish and chips as:
       (1) a fraction   (2) a decimal   (3) a percentage
   b A new customer comes in to buy a meal.
       What meal is the customer most likely to choose?

4 Leslie spins this spinner and rolls this dice.
   a Draw a sample space diagram to show all
       the possible outcomes.
   b Use your diagram to write down the
       probability of getting:
       (1) red and a 6   (3) blue and a number
       (2) green and a 1   greater than 4

5 A 20 p piece and a 5 p piece are tossed together.
   What is the probability of getting two heads?

6 A coin is tossed, then a card is chosen
   from a pack of 52 playing cards.
   a Complete the tree diagram.
   b What is the probability of
       getting a head and a heart?

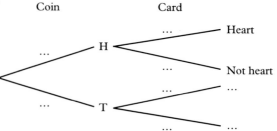

12 Loci: know your place

In maths an ellipse is defined as the locus of a point which moves so that the sum of its distances from two fixed points remains constant.

To draw an ellipse you need two nails fixed to paper on a board, a piece of string tied into a loop and a pencil. Keep the string taut and move the pencil around the two nails until you end up where you started.

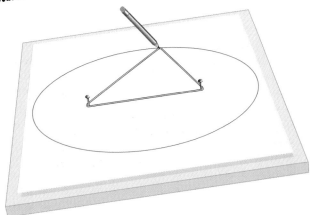

1 Locus of a point

Exercise 12:1

You will need a sheet of paper and some counters.

1 a Mark a point in the centre of your paper.
Label it A.

b Put a counter on the paper 7 cm from A.
Put some more counters on your paper so
that they are also 7 cm from A.

c All your counters should lie on a curve.
Describe this curve.

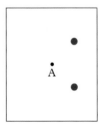

2 a Mark two points on your paper.
Label them B and C.

b Put a counter on the paper so that it is the
same distance from both B and C.
Put more counters on the paper so that they
are all the same distance from both B and C.

c All your counters should lie on a straight line.
Describe the position of this line.

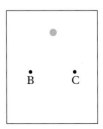

3 **a** Draw a line 7 cm long in the middle of your paper.
  b Put a counter on the paper so that it is 5 cm from the line.
   Put more counters on the paper so that they are all 5 cm from the line.
  c Describe where your counters can be.
   This should include some straight parts and some curved parts.

Locus	The position of an object can be given by a rule.

The **locus** is all the possible positions that the object can take that satisfy the rule.
The locus can either be described in words or by drawing.

Example   Describe the locus of an object that is always 2 cm from a point D.

The locus is a circle. The centre is D and the radius is 2 cm.

---

4 **a** Describe in words the locus of an object that is always 4 cm from a point R.
  b Sketch the locus of the object.

5 **a** Copy this line on to paper.

A ————————————————— B

  b Draw a sketch of the locus of a point that is always 8 cm away from the line AB.

6 Describe the locus of the tip of the minute hand as it moves around the clock face.

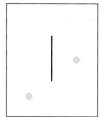

7 Describe the locus of the tip of the arrow as it moves over the scale of this ammeter.

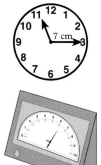

Example   John's house is 2 km from the school and 3 km from the motorway.
Where could John's house be?

School

Motorway

John's house is 2 km from the school.
His house lies on a circle whose centre is the school.
The radius is 2 km.

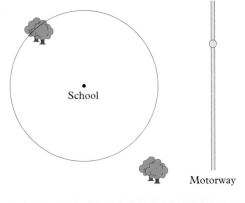

School

Motorway

The house is 3 km from the motorway.
It lies on a line 3 km from the motorway.
This line is parallel to the motorway.

John's house must be at one of the points where the circle and the line cross.
These are marked with a cross

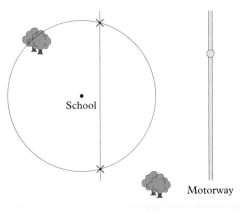

School

Motorway

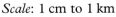

Scale: 1 cm to 1 km

Exercise 12:2

Use tracing paper to copy these maps.
The scale of each map is 1 cm to 1 km.

1 Katy's house is 3 km from school and
4 km from the motorway.
  a Copy the diagram.
  b Find the two possible positions of
    Katy's house.

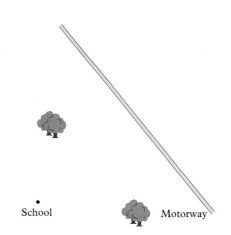

School   Motorway

2 Brian lives 1 km from the motorway
and 2 km from the edge of the forest.
  a Copy the diagram.
  b Find the position of Brian's house.

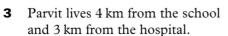

Motorway

3 Parvit lives 4 km from the school
and 3 km from the hospital.
  a Copy the diagram.
  b Find the two possible positions
    of Parvit's house.

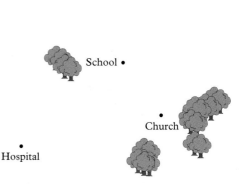

School •

Church

Hospital

4 Sam lives 2 km from the school and 5 km from the hospital.
He lives less than 1 km from the church.
  a Make another copy of the map in question **3**.
  b Mark the only possible position of Sam's home.

Example

A goat is tethered to a ring in the ground.
The rope is 3 m long.
Draw a diagram to show the area where the goat can graze.

The point R is the ring.
The red shading shows the area where the goat can graze.
The radius of the circle is the length of the rope.

3 m
R

You should always say which part of the shading is your answer.
A key is useful when there are two or more types of shading.

Exercise 12:3

1 The horse is tied to a wall as shown.
The length of the rope is 4 m.
Make a sketch showing the area where the horse can reach.

2 This bull is tethered to a ring in the farmyard.
The rope is 5 m long.
Make a sketch showing the area where the bull can reach.

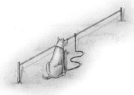

W You will need worksheet 12:1 for the rest of this exercise.

3 The guard dog is on a chain of length 3 m.
The other end of the chain is fixed to a ring.
The ring can move along the rail.
The rail is 7 m long.
Use shading to show the area where the dog can reach.

4 A and B show the positions of
two radio transmitters.
Each transmitter can cover a
distance of 40 km.

  a Draw the locus of the area
covered by transmitter A.

  b Draw the locus of the area
covered by transmitter B.

  c Use shading and a key to show
the area covered by both
transmitters.

● A   ● B

Scale: 1 cm to 10 km

5 The diagram shows the floor
area of a room. The points A
and B show the positions of
alarm sensors. Each sensor
can cover a distance of up
to 4 m.

  a Draw the locus of the area
covered by sensor A.

  b Draw the locus of the area
covered by sensor B.

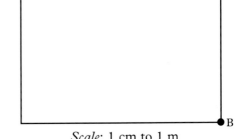

Scale: 1 cm to 1 m

  c Use shading and a key to show the area covered by both sensors.

  d Draw a new diagram showing the floor area of a room 7 m by 4 m.
Use the same sensors.
Shade the floor area not covered by either of the two sensors.

6 Laura plugs her electric
mower into the socket on the
side of the house.
The mower lead is 5 m long.

  a Shade the part of the lawn that
Laura can reach with the
mower.

  b What is the minimum length
of lead that Laura would need
to cover all the lawn?

Scale: 1 cm to 1 m

Example

Kay wants to sketch the locus of the valve on the rim of a bicycle wheel as the bike moves forward.

Kay uses a piece of paper and a circle of card to help her.
She makes a mark on the edge of the circle.

Kay rolls the circle along the ruler bit by bit.
Each time, she makes a mark on the paper beside the mark on the circle.

Kay joins up the marks.

The curve is the locus of a point on a wheel as it moves forward.

Exercise 12:4

1 Patrick has put a mark on
the corner of this box.
He finds the locus of this
corner as the box rolls over.

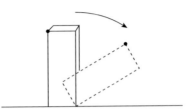

Use a ruler and a rectangular
piece of card.
Mark a corner of the card.
Roll the card along the ruler. Mark the paper as the box rolls bit by bit.
Join up your marks with a curve.

2 **a** Use a ruler and a circle made out of card like Kay's.
Make a sketch of the locus of a point on the circle as the circle
moves forward.

b Find the locus of the corner
of this square as it rolls forward.

c Investigate what happens with
other regular shapes.

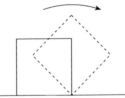

3 Rolf rolls a coin around the inside
edges of a box.
Sketch the locus of the centre of
the coin as it moves around the box.

4 Mark puts his ladder against the house.
The ground is slippery.
The ladder slides down until it is lying
on the ground.
Draw the locus of the centre of the
ladder as it slides down the wall.

Hint: Use the edges of a piece of paper
to be the wall and the ground.
Use a ruler for the ladder.

2 **Constructions**

This aircraft is flying exactly
down the middle of two cliffs.
It keeps exactly between the two
cliffs to stay on course.

Bisecting an angle	**Bisecting an angle** means splitting it exactly in half. You do not need an angle measurer to do this. It is more accurate to do it with compasses.

Exercise 12:5

1 a Draw a 60° angle.

b Open your compasses a small distance.
You **must** keep your compasses fixed
from now on.

c With the compass point on the corner of
the angle, draw a small arc which crosses
both arms of the angle. Label these points
A and B.

d Place your compasses on point A and
draw another arc in the middle of the
angle.

e Now place your compasses on point B.
Draw another arc in the middle of the
angle.
It should cross the first one.

f Finally, draw a line from the corner of the
angle through the point where your two
arcs cross.
This line bisects the angle.

g Measure the two parts of the angle to
check that it is correct.

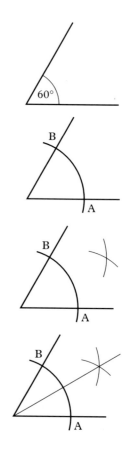

2 **a** Draw a 45° angle.
  b With your compass point at the corner of the angle,
     draw an arc crossing both arms .
     Label the crossing points A and B.
  c Draw arcs from points A and B.
  d Draw the line which bisects the angle.
  e Measure the two parts of the angle to check that it is correct.

3 By bisecting an angle, draw an angle of 45°.

4 **a** Draw an angle of 120°.
  b Bisect the angle.
  c Bisect the resulting angle. Check that this new angle is 30°.

Equidistant from two lines   All the points on the line bisecting an angle are **equidistant** from the two arms of the angle.
This means that they are always the same distance from the two lines. Both the blue lines are the same length. So are both the red ones.

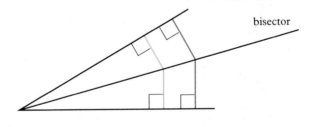

---

5 **a** Draw two lines AB and AC which are at 48° to each other.
  b Draw the locus of the points which are equidistant from AB and AC.

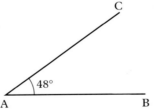

6 **a** Draw two lines PQ and PR which are at 130° to each other.
  b Draw the locus of the points which are equidistant from PQ and PR.

Exercise 12:6

1 a Draw a horizontal line 8 cm long.
Leave some space above it.
Label the ends of the line A and B.

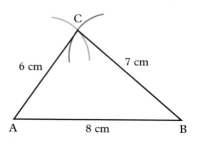

b Set your compasses to 6 cm.
Draw an arc, centre A (blue arc).

c Set your compasses to 7 cm.
Draw an arc, centre B (red arc).

d Join A and B to the point where
your arcs cross.
Label this point C. You should now have a triangle.

e Bisect the angle at vertex (corner) A of the triangle.

f Bisect the angle at vertex B of the triangle.

g Bisect the angle at vertex C of the triangle.
The three bisectors should cross at one point.

2 a Construct an equilateral triangle with
sides 10 cm.

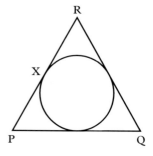

b Label your triangle PQR.

c Bisect all three angles of the triangle.

d Mark the point X at the middle of PR.

e Place the point of your compasses on
the point where the bisectors cross.
Carefully move your pencil point until
it touches point X.

f Draw a circle with your compasses.
This circle should just touch all three sides of the triangle.
This is called the **inscribed circle**.

3 A wall and a fence are at 50° to each other.
Shade the area which is nearer to the wall than the fence.

4 A small rectangular flag measures 10 cm by 5 cm.
Its design is created by bisecting each angle and extending the bisector
to the edge of the rectangle.
Make an accurate drawing of the flag and colour it as you wish.

Bisecting a line	**Bisecting a line** means cutting it exactly in half.

Perpendicular bisector	Two lines which are at right angles are called perpendicular. On this diagram, CD is perpendicular to AB and CD bisects AB. CD is called the **perpendicular bisector** of AB.

Exercise 12:7

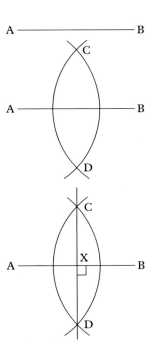

1 a Draw a line 10 cm long.
Label the ends A and B.
b Put your compass point on A.
Move your pencil until you can tell it is more than half way along the line.
c Draw an arc from above the line to below it.
d **Without changing your compasses,** move the compass point to B.
e Draw another arc from B.
f Your arcs should cross above and below the line.
Label these points C and D.
g Join C to D. Use a ruler.
Line CD bisects line AB at right angles.
Label the point where CD and AB cross.
Call it X.
Measure AX and BX with a ruler to check that they are equal.

2 a Draw a line 8.5 cm long.
b Label your line AB.
c Draw arcs from A and B.
d Label the crossing points of the arcs C and D.
e Join C to D to bisect the line at right angles.
f Check that CD bisects AB by measuring.

3 a Draw a line 6.4 cm long.
b Label the line AB.
c Bisect the line using a ruler and compasses.

Equidistant from two points

All the points on the perpendicular bisector of a line are **equidistant from the two points** at the ends of the original line.
Every point on CD is the same distance from A and B.
You can describe CD as the locus of the points that are equidistant from A and B.

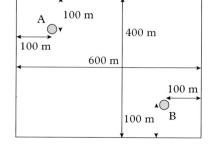

4 **a** Draw points A and B which are 7 cm apart.
  b Join the points with a straight line.
  c Draw the locus of the points which are equidistant from A and B.

5 The diagram shows a plan of a rectangular park.
A and B are two drinking fountains.
  a Using a scale of 1 cm to 50 m, draw a scale plan of the park.
  b Draw a line on your diagram to help you show the areas of the park which are nearer to fountain A.

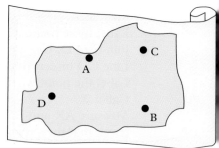

6 **a** Make a rough copy of this treasure map. It does not have to be exact.
  b The treasure is buried at a point which is equidistant from A and B. It is also equidistant from C and D.
    Find the treasure!

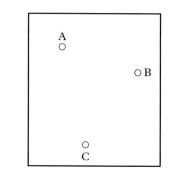

7 The diagram shows a large room and the position of three TV monitors.
  a Make a rough copy of the diagram.
  b Lightly shade the area which is nearer to monitor A than to monitor B.
  c Shade the area which is nearer to monitor A than to monitor C.
  d Show the area which is closer to monitor A than either of the other two monitors.

Exercise 12:8

1 **a** Construct a triangle with sides 10 cm, 8 cm
and 7 cm.
Start by drawing a horizontal line 10 cm long.

   b Label your triangle PQR.

   c Draw the perpendicular bisectors of all
three sides of the triangle.

   d Place the point of your compasses on the
point where the bisectors cross.
Carefully move your pencil point until it
touches point P.

   e Draw a circle with your compasses.
This circle should just touch all three points P, Q and R.
This is called the **circumscribed circle** or the **circumcircle**.

2 **a** Construct a triangle with sides 9 cm, 7 cm and 8 cm.
Start by drawing a horizontal line 9 cm long.
Label your triangle PQR.

   b Draw the perpendicular bisectors of all three sides of the triangle.
Label the point where the bisectors cross X.

   c Draw the circumcircle of the triangle.

3 **a** Draw a rectangle 12 cm by 8 cm.

   b Draw in the diagonals.

   c Draw the perpendicular bisector of each diagonal.

   d Colour in the pattern you have made.

4 **a** Draw a circle of radius 6 cm.

   b Mark 4 points W, X, Y and Z
around the circumference of the
circle.

   c Draw in the lines WY and XZ.

   d Draw the perpendicular bisectors
of WY and XZ.
The crossing point of these
bisectors should be the centre of
the circle.

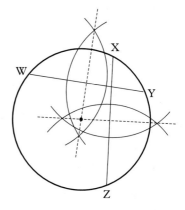

1 Describe the locus of each of these:
a a door handle as the door opens,
b the foot of a person running,
c a chair on this fairground ride.

2 This dog is chained to a ring in the wall.
The length of the chain is 3 m.
Make a sketch showing the area where the dog can reach.

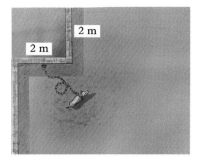

3 Keith is rowing across a river 30 m wide.
There is a strong current downstream.
For each complete stroke of the oars Keith moves 2 m across and 1 m down the river.
a Draw a line on your paper to represent the bank of the river.
b Mark Keith's starting point.
c Mark his position after one stroke of the oars.
d Do this for three more strokes of the oars.
e How far down the river will Keith have moved by the time he reaches the opposite bank?

Flow of river

Start

Scale: 1 cm to 2 m

4 **a** Draw a 70° angle.
  b With your compass point on the corner, draw an arc crossing both
    arms of the angle.
    Label the crossing points A and B.
  c Draw arcs from points A and B.
  d Draw the line which bisects the angle.
  e Measure the two parts of the angle to check that it is correct.

5 **a** Draw two lines AB and AC which are at 64° to each other.
  b Draw the locus of the points which are equidistant from AB
    and AC.

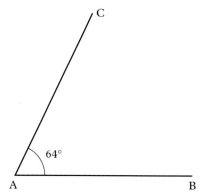

6 **a** Draw this triangle accurately.
  b Bisect the angle at vertex A.
  c Bisect the angle at vertex B.
  d Bisect the angle at vertex C.
    The three bisectors should
    cross at one point.

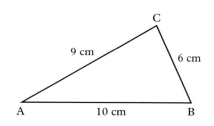

7 **a** Draw a line 5 cm long.
  b Label the end points A and B.
  c Draw the locus of the points which are equidistant from A and B.

8 It is possible to draw a circumcircle on any triangle.
  The centre of the circle can be either inside the triangle or on one of
  the sides of the triangle or outside the triangle.
  Investigate the position of the centre of the circle for different triangles.
  When will the centre of the circle be outside the triangle?

1 **a** Make a copy of this diagram.
   b Shade the locus of all points
      2 cm or less from A.
   c Shade the locus of all points
      3 cm or less from B.
   d Shade the locus of all points
      2 cm or more from C.
   e Mark the points that satisfy
      all three loci.
      (Loci is the plural of the
      word locus.)

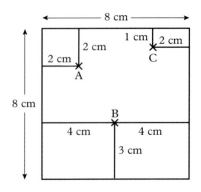

2 **a** Draw a circle of radius 8 cm.
   b Mark two points P and Q anywhere inside the circle.
      Join P to Q with a straight line.
   c Find a point on the circumference of the circle which is equidistant
      from P and Q.
   d Find any other points which are also equidistant from P and Q.

3 This diagram shows the positions of Alan, Barbara and Charlie.
   They all started at the same point and ran away from each other at the
   same speed.
   Find the point where they started.

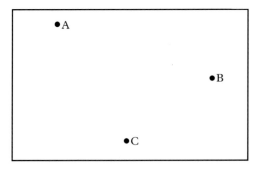

4 **a** Draw a line AB of length 8 cm.
   b Draw the locus of all points C so that the triangle ABC has an area
      of 24 cm².

● **Locus**
The position of an object can be given by a rule.
The **locus** is all the possible positions that the object can take that satisfy the rule.
The locus can either be described in words or by drawing.

Example
Describe the locus of an object that is always 2 cm from a point D.

The locus is a circle. The centre is D and the radius is 2 cm.

● **Equidistant from two lines**
All the points on the line bisecting an angle are **equidistant** from the two arms of the angle.
This means that they are always the same distance from the two lines. Both the blue lines are the same length. So are both the red ones.

● **Bisecting a line**
Bisecting a line means cutting it exactly in half.

Perpendicular bisector
Two lines which are at right angles are called perpendicular.
On this diagram, CD is perpendicular to AB and CD bisects AB.
CD is called the **perpendicular bisector** of AB.

● **Equidistant from two points**
All the points on the perpendicular bisector of a line are **equidistant from the two points** at the ends of the original line.

Every point on CD is the same distance from A and B.

You can describe CD as the locus of the points that are equidistant from A and B.

1 Pat's house is 4 km from the school
and 3 km from the motorway.
  a Copy the diagram.
  b Find the two possible positions
     of Pat's house.

School

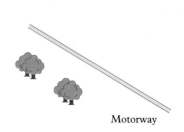

Motorway

Scale: 1 cm to 1 km

2 Don is watering his lawn.
He has two sprinklers, A and B.
They both spray areas up to 4 m
away. The diagram shows Don's
lawn and the sprinklers.
  a Make a sketch of Don's lawn.
  b Show the area of lawn that is
     watered by both sprinklers.

Scale: 1 cm to 2 m

3 Janine wants to put a path across this field.
She wants the footpath to cross the field
so that it is the same distance from both
sides of the field.
  a Make a copy of the diagram using
     an angle of 70° between the two sides.
  b Use a construction to show where
     the path should go.

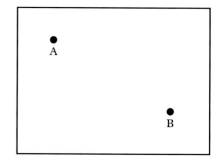

4 **a** Draw points P and S which are 8 cm apart.
  b Join the points with a straight line.
  c Draw the locus of the points which are equidistant from P and S.

13 Algebra: two at a time

These two discs each have a different number on their reverse side.

Adding any pair of the four numbers gives these totals:

27, 23, 32, 18

Find the two numbers.

1 ◀◀ REPLAY ▶

The gradient tells you how steep the hill is.

We use the word gradient in mathematics to describe the steepness of straight lines.

Equation	The rule that describes a line is called the **equation** of the line. This line has the rule: y co-ordinate = x co-ordinate + 2 The equation is $y = x + 2$

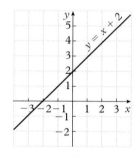

Exercise 13:1

1
  a Copy the axes on to squared paper.

  b Write down the co-ordinates of three points where y co-ordinate = x co-ordinate
  Plot these points.

  c Join the points with a straight line.
  Label the line with its equation $y = x$

  d Write down the co-ordinates of three points where y co-ordinate = $2 \times x$ co-ordinate
  Plot these points on the same set of axes. Join them with a straight line.
  Label the line $y = 2x$

  e On the same set of axes draw and label the lines $y = 3x$ $y = \frac{1}{2}x$

  f Which part of the equation tells you how steep the line is?

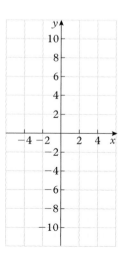

2 Which line is steeper in each of these pairs?

a $y = 4x$ or $y = 8x$

b $y = 7x$ or $y = 2x$

c $y = \dfrac{x}{3}$ or $y = x$

d $y = \dfrac{x}{5}$ or $y = \dfrac{x}{4}$

3 **a** Draw a set of axes on squared paper.
Use values of x from -4 to 4 and values of y from -4 to 8.

b Copy and complete
this table for $y = -x$
Draw and label the line $y = -x$

x	-1	0	1	2
y	1	0		

c Copy and complete
this table for $y = -2x$
Draw and label the line $y = -2x$

x	-1	0	1	2
y		0	-2	

d Draw and label the line $y = -3x$

e Look at the three graphs you have drawn.
(1) Which line is the steepest?
(2) Which line is the least steep?

4 **a** Copy the axes on to squared paper.

b Draw and label these lines
$y = x$ $y = x + 1$ $y = x + 3$

c Draw and label these lines
$y = x - 1$ $y = x - 2$ $y = x - 3$

d What do you notice about all the lines?

e Which part of the equation tells you
where the line will cross the y axis?

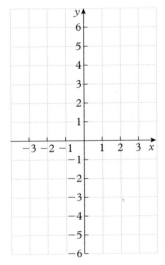

5 Which line is steeper in each of these pairs?

   a $y = -3x + 5$ or $y = -2x - 4$

   b $y = -2x - 1$ or $y = 1 - 5x$

   c $y = -2x + 6$ or $y = 8 - 3x$

   d $y = 3x + 2$ or $y = -4x + 2$

6 Where will these lines cross the y axis?

   a $y = 3x + 7$ **d** $y = 7x - \frac{3}{4}$

   b $y = 2x - 5$ **e** $y = 5 - 2x$

   c $y = -4x + \frac{1}{2}$ **f** $y = -5x$

Gradient

The **gradient** of a line tells you how steep the line is.

Both these lines have a '3' to tell you how steep they are.

They have **gradient** 3. The lines are parallel because they have the same gradient.

The $+2$ and the -4 tell you where the lines cross the y axis.

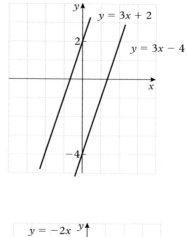

Both these lines have gradient -2.

They are parallel.

They cross the y axis at 0 and -1.

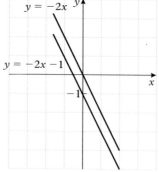

---

7 Look at the equation $y = mx + c$

It is the equation of a straight line.

   a Write down the gradient of the line.

   b Write down the co-ordinates of the point where the line crosses the y axis.

Example

Find the equation of the red line.
The red line is parallel to the line
$y = 2x + 2$
Part of the equation must be $y = 2x$ ...
The red line crosses the y axis at -3.
The equation of the red line must
be $y = 2x - 3$

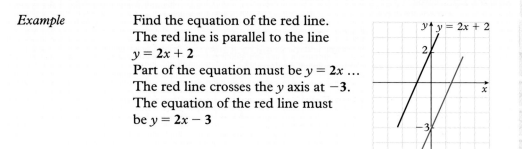

Exercise 13:2

Write down the equation of each red line.

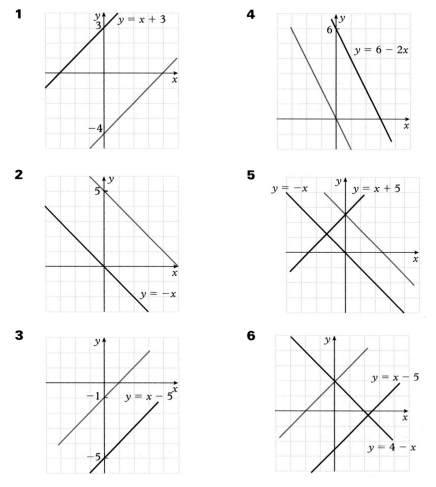

1

$y = x + 3$

2

$y = -x$

3

$y = x - 5$

4

$y = 6 - 2x$

5

$y = -x$ $y = x + 5$

6

$y = x - 5$

$y = 4 - x$

Exercise 13:3

1 **a** Draw a pair of axes on squared paper.
  Use values of x from -4 to 4 and values of y from -9 to 6.
  b Draw and label the line $y = 2x - 3$
  c Is the point $(-4, -7)$ above or below the line?
  d These points lie on the line. Fill in the missing co-ordinates.
    $(..., -5)$ $(-2, ...)$ $(..., -3)$ $(..., 0)$

2 This is the graph of $y = \frac{1}{2}x - 1$
  Use the graph to answer these questions.
  a What is the value of y when $x = 4$?
  b What is the value of x when $y = -3$?
  c Which of these points lie below the
    line?
    $(2, -1)$ $(3, 3)$ $(1, 2)$
  d These points lie on the line.
    Fill in the missing co-ordinates.
    $(..., -2)$ $(0, ...)$

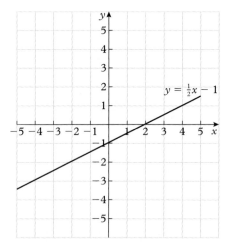

$y = \frac{1}{2}x - 1$

Example Does the point $(2, 5)$ lie on the line $y = x + 4$?

When $x = 2$ $y = 2 + 4$
                      $= 6$

The point $(2, 6)$ lies on the line.
The point $(2, 5)$ does not lie on the line.

---

3 Each of parts **a** to **e** has a point and a line.
  Does the point lie on the line?
  a $(3, 7)$ $y = x + 4$
  b $(2, 6)$ $y = 2x + 3$
  c $(1, 6)$ $y = 3x - 1$
  d $(0, -5)$ $y = 2x - 5$
  e $(4, 16)$ $y = 5x - 3$

The equations of lines are not always written in this form.

Example

Draw the line $2x + y = 6$

There is a quick method to
draw a line like this.
Find the points when
$x = 0$ and $y = 0$

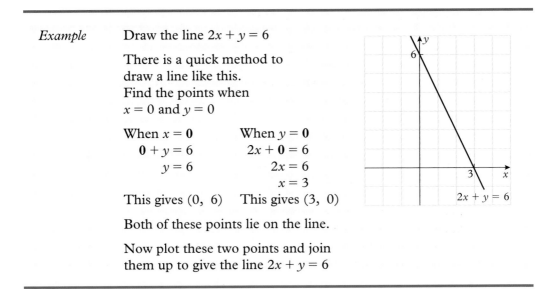

When $x = 0$	When $y = 0$
$0 + y = 6$	$2x + 0 = 6$
$y = 6$	$2x = 6$
	$x = 3$
This gives $(0, 6)$	This gives $(3, 0)$

Both of these points lie on the line.

Now plot these two points and join
them up to give the line $2x + y = 6$

Exercise 13:4

1 Copy the axes on to squared paper.
   a Find the co-ordinates of two
      points on the line $x + y = 3$
   b Plot the points.
      Draw and label the line
      $x + y = 3$

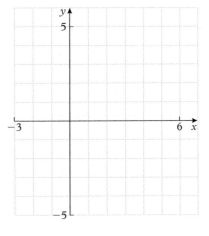

2 a Find the co-ordinates of two points on the line $x + 2y = 6$
   b Plot the points on the same set of axes as question **1**.
      Draw and label the line $x + 2y = 6$
   c Find the co-ordinates of two points on the line $2x + 3y = 12$
      Plot the points.
      Draw and label the line $2x + 3y = 12$

2 Simultaneous equations

It is easy to see where these two lines of gymnasts cross at the box.
In mathematics you have to find where two lines cross.
You can do this by drawing lines or by using algebra.

| **Point of intersection** | The point where two lines cross is called the **point of intersection**. |

Example

Alex and Tom are buying some food at the youth club.
Alex buys two *b*iscuits and one *d*rink for 10 p.
Tom buys one *b*iscuit and two *d*rinks for 14 p.
Find the cost of a *b*iscuit and the cost of a *d*rink.

Alex's equation is $2b + d = 10$
When $b = 0$ $\qquad d = 10$
When $d = 0$ $\qquad 2b = 10$
$\qquad\qquad\qquad b = 5$
Two points on this line
are $(0, 10)$ and $(5, 0)$.

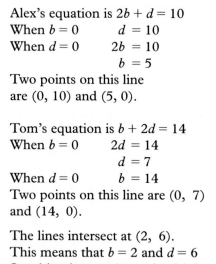

Tom's equation is $b + 2d = 14$
When $b = 0$ $\qquad 2d = 14$
$\qquad\qquad\qquad d = 7$
When $d = 0$ $\qquad b = 14$
Two points on this line are $(0, 7)$
and $(14, 0)$.

The lines intersect at $(2, 6)$.
This means that $b = 2$ and $d = 6$
So a biscuit costs 2 p and a drink costs 6 p.

Check: $2b + d = 2 \times 2 + 6 = 10$ ✓ $\qquad b + 2d = 2 + 2 \times 6 = 14$ ✓

Simultaneous equations   When you solve two equations at the same time you are solving **simultaneous equations**.

Exercise 13:5

Draw graphs to solve these problems.
Check your answers in the original problem each time.

1 Solve these pairs of simultaneous equations.

    a $x + y = 5$
           $2x + 4y = 12$

    ● **b** $x + y = 7$
              $3x + y = 11$

2 John and Alisha took part in a school quiz. They had to choose a *s*tandard or a *h*ard question on each turn.
John answered 3 *s*tandard and 2 *h*ard questions correctly and scored 12 points.
Alisha answered 1 *s*tandard and 4 *h*ard questions correctly and scored 14 points.
Find the points awarded for a *s*tandard question and for a *h*ard question.

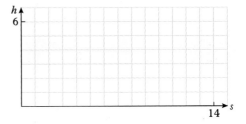

3 A school sells two types of calculator. One is a *b*asic model and the other is a *s*cientific model.
The cost of one *b*asic and one *s*cientific calculator is £10.
The cost of 3 *b*asic and 2 *s*cientific calculators is £24.
Find the cost of a *b*asic model and the cost of a *s*cientific model.

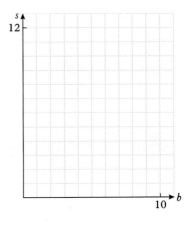

You can also solve simultaneous equations using algebra.

Example

Solve this pair of simultaneous equations $\quad 5x + y = 20$
$\qquad\qquad\qquad\qquad\qquad\qquad\qquad\qquad\qquad\qquad 2x + y = 11$

Number the equations

(1) $\ 5x + y = 20$
(2) $\ 2x + y = 11$

Subtract to get rid of y
This finds x

$3x\quad = 9$
$x = 3$

Use equation (1) to find y

Put $x = 3$ in equation (1)
$5 \times 3 + y = 20$
$15 + y = 20$
$y = 5$
The answer is $x = 3$, $y = 5$

Use equation (2) to check your answer
$2x + y = 2 \times 3 + 5 = 6 + 5 = 11$ ✓

Exercise 13:6

Solve these pairs of simultaneous equations.
Start by subtracting the equations each time.

1 $\quad 5x + y = 13$
$\qquad x + y = 5$

5 $\quad 5x + y = 23$
$\qquad 2x + y = 14$

2 $\quad 3x + y = 22$
$\qquad x + y = 12$

6 $\quad 3r + 2s = 16$
$\qquad r + 2s = 12$

3 $\quad 5a + b = 28$
$\qquad 2a + b = 13$

7 $\quad 4x + 3y = 25$
$\qquad x + 3y = 13$

4 $\quad 7x + y = 18$
$\qquad 3x + y = 10$

8 $\quad 5d + 2f = 19$
$\qquad d + 2f = 15$

Example

Solve this pair of simultaneous equations
$$3x + y = 19$$
$$x - y = 1$$

Number the equations

(1) $3x + y = 19$
(2) $x - y = 1$

Add to get rid of y
This finds x

$4x = 20$
$x = 5$

Use equation (1) to find y

Put $x = 5$ in equation (1)
$$3 \times 5 + y = 19$$
$$15 + y = 19$$
$$y = 4$$
The answer is $x = 5$, $y = 4$

Use equation (2) to check your answer
$x - y = 5 - 4 = 1$ ✓

Exercise 13:7

Solve these pairs of simultaneous equations.
Start by adding the equations each time.

1 $2x + y = 8$
$x - y = 7$

4 $4x + y = 12$
$3x - y = 2$

2 $4x + y = 7$
$5x - y = 2$

5 $3x + 2y = 17$
$5x - 2y = 7$

3 $3x + y = 18$
$x - y = 2$

6 $4x + 3y = 26$
$5x - 3y = 19$

Exercise 13:8

Solve these pairs of simultaneous equations.
You need to decide whether to add or subtract the equations.

1 $2x + y = 11$
$3x - y = 9$

4 $4x + 2y = 38$
$3x - 2y = 11$

2 $3x + 2y = 16$
$x + 2y = 12$

5 $5x + 3y = 19$
$2x + 3y = 13$

3 $4p + 2q = 28$
$p - 2q = 2$

6 $5a + 3b = 27$
$4a - 3b = 0$

You sometimes have to multiply one of the equations before adding or subtracting.

Example   Solve this pair of simultaneous equations $2x + 3y = 13$
$4x - y = 5$

Number the equations   (1) $2x + 3y = 13$
(2) $4x - y = 5$

You need to multiply equation (2) by **3**
so that you have $3y$ in $2x + 3y = 13$
each equation   (2) $\times$ **3** $12x - 3y = 15$

Add to get rid of y $14x = 28$
This finds x $x = 2$

Use equation (1) to find y Put $x = 2$ in equation (1)
$2 \times 2 + 3y = 13$
$4 + 3y = 13$
$3y = 9$
$y = 3$
The answer is $x = 2$, $y = 3$

Use equation (2) to check your answer
$4x - y = 4 \times 2 - 3 = 8 - 3 = 5$ ✓

Exercise 13:9

Solve these pairs of simultaneous equations.
You will need to multiply one equation by a number.

1 $3x + 2y = 16$
$x + y = 7$

4 $5x + 3y = 36$
$x + y = 10$

2 $4x - 2y = 6$
$3x + y = 17$

5 $3x + 2y = 9$
$4x - y = 1$

3 $7a - 3b = 17$
$2a + b = 16$

● **6** $f + 4g = 7$
$5f - 2g = 26$

Sometimes both equations have to be multiplied before adding or subtracting.

Example

Solve this pair of simultaneous equations
$$3x + 5y = 30$$
$$2x + 3y = 19$$

Number the equations

(1) $3x + 5y = 30$
(2) $2x + 3y = 19$

Multiply equation (1) by 2
Multiply equation (2) by 3

$6x + 10y = 60$
$6x + 9y = 57$

You can now subtract to get rid of x
Put $y = 3$ in equation (1)

$y = 3$
$3x + 15 = 30$
$3x = 15$
$x = 5$

The answer is $x = 5, \ y = 3$

Check using equation (2)
$2x + 3y = 2 \times 5 + 3 \times 3 = 10 + 9 = 19$ ✓

Exercise 13:10

Solve these pairs of simultaneous equations.
You will need to multiply both equations.

1 $2x + 3y = 11$
$5x + 4y = 24$

5 $4x + 3y = 12$
$5x + 7y = 15$

2 $3x - 2y = 13$
$4x + 3y = 40$

6 $7p - 3q = 5$
$3p + 2q = 12$

3 $7a + 4b = 41$
$2a + 5b = 31$

7 $5x - 3y = 24$
$3x - 2y = 14$

4 $3x + 2y = 33$
$7x + 5y = 79$

8 $2c - 3d = 6$
$5c - 7d = 17$

Example   Nathan has been given these   (1) $2x = 2 + y$
equations to solve simultaneously.   (2) $3x + 4y - 36 = 0$

He has to change the equations to get them in the right form.

Subtract y from both sides of equation (1) $2x - y = 2$
Add 36 to both sides of equation (2) $3x + 4y = 36$

Nathan can now solve the equations (1) × 4 $8x - 4y = 8$
                                                   $3x + 4y = 36$
                                                   _____
                        Adding $11x\ \ \ \ = 44$
                                                   $x = 4$
            Put $x = 4$ in (2) $3x + 4y = 36$
                                    $12 + 4y = 36$
                                          $4y = 24$
                                            $y = 6$
        The answer is $x = 4$ and $y = 6$

Check using equation (1)
$2x - y = 2 \times 4 - 6 = 2$ ✓

Exercise 13:11

Solve these pairs of simultaneous equations.
You will need to change the equations to get them in the right form.

1 $x = 2 + y$
        $3x + y = 14$

2 $2a = 3b + 1$
        $a + 2 = 2b$

3 $x = 10 - y$
        $y = x + 8$

4 $2p = 10 - q$
        $4q = 1 + 5p$

5 $2x = y + 3$
        $2 = y - x$

6 $3a = 16 - 2b$
        $3b = 23 - 4a$

7 $f = g + 5$
        $3g = f + 1$

8 $3x - 4y - 5 = 0$
        $26 = 2x + 3y$

3 Inequalities

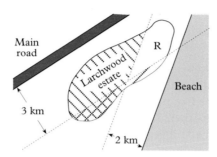

Main road

Larchwood estate

R

Beach

3 km

2 km

The James family are looking for a new house.

They want to live less than 2 km from the beach and less than 3 km from the main road.

They have crossed out the parts of the estate where they do not want to live.

R is the region of the estate where they will look for a house.

Example

a Write down the equation of the vertical line.

b Write down the co-ordinates of the point of intersection of the two lines.

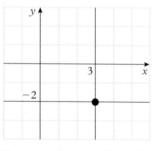

a All points on the vertical line have x co-ordinates of *3*. The equation of the line is $x = 3$

b The co-ordinates of the point of intersection are $(3, -2)$.

Exercise 13:12

1 a Write down the rules for each of these lines.

Write down the co-ordinates of the points of intersection of these pairs of lines.

b E and C

c A and D

d B and F

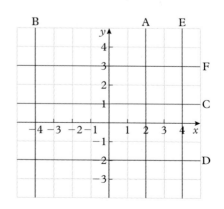

Examples

1 Find the region where
$x \le 2$ for all points.

The region includes all
points whose x co-ordinate
is less than or equal to 2.

Draw the line $x = 2$
All points to the left of
the line have an x co-ordinate
less than 2.
Shade this region.

The required region is shaded.

2 Find the region where
$y > 1$ for all points.

The region includes all
points whose y co-ordinate
is greater than 1.

Draw the line $y = 1$
Make it a dotted line because
$y = 1$ is not included in $y > 1$.

All points above the line have
a y co-ordinate greater than 1.
Shade this region.

The required region is shaded.

Exercise 13:13

Sketch the regions where these inequalities are satisfied.
Use a separate diagram for each one.

1 $x \ge 1$

2 $x < 4$

3 $y \le 3$

4 $y > 2$

5 $x \ge -3$

6 $y > -1$

7 $y \le 5$

8 $x \le -2$

9 $y < -3$

Example   Find the region where
$y < x + 2$ for all points.

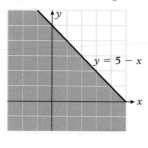

The region includes all points
whose y co-ordinate is less than
the value of the x co-ordinate + 2.
Draw the line $y = x + 2$
Make it a dotted line because
$y = x + 2$ is not included in $y < x + 2$
All points below the line have a y co-ordinate less than $x + 2$.
Shade this region.

If you are not sure which side of the line to shade, use this method
to check:
Choose a point on one side of the line, for example (3, 1)
Put the values of x and y in the inequality $y < x + 2$
$$1 < 3 + 2$$

If the equality is still true you have chosen the correct side.

Exercise 13:14

Make a separate sketch to show each of these regions.

1 $y > 2x$ **3** $y \geqslant x - 4$ **5** $y \geqslant 4 - x$

2 $y \leqslant x + 3$ **4** $y < 3x - 1$ **6** $y > 4 - 2x$

7 Write down the inequality for the shaded region in each diagram.

a

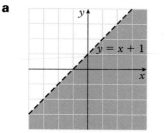

c

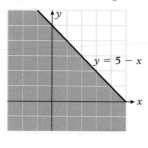

b

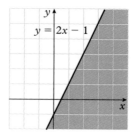

d

Example   Find the region that is described by these inequalities.
$$y \leqslant 4 \qquad x > -1 \qquad y \geqslant x + 1$$

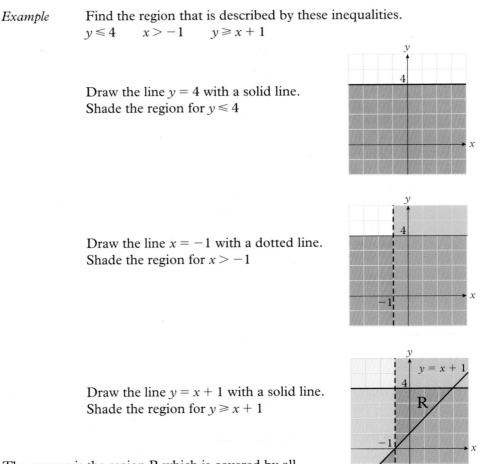

Draw the line $y = 4$ with a solid line.
Shade the region for $y \leqslant 4$

Draw the line $x = -1$ with a dotted line.
Shade the region for $x > -1$

Draw the line $y = x + 1$ with a solid line.
Shade the region for $y \geqslant x + 1$

The answer is the region R which is covered by all three shadings.

Exercise 13:15

Find the region that is described by each of these sets of inequalities.

1 $y \geqslant -3$ $x \leqslant 2$ $y \leqslant x$

2 $x > -2$ $y \leqslant 4$ $y \geqslant x$

3 $y \leqslant 5$ $x > -2$ $y \geqslant 2x - 3$

4 $x > 0$ $y > -2$ $y \leqslant 1 - x$

5 $x \geqslant -3$ $x < 2$ $y \leqslant 5$ $y > -1$

6 $1 < x \leqslant 4$ $-2 \leqslant y < 3$

1 **a** Copy and complete this table for $y = 2x - 1$

x	0	1	2
y	...	...	...

   b Draw a set of axes on squared paper.
      Use values of x from -3 to 3 and values of y from -7 to 5.
   c Draw and label the line $y = 2x - 1$
   d Is the point (2, 1) above or below the line?
   e The point (..., -3) lies on the line.
      What is the missing x co-ordinate?
   f What is the value of y when $x = -2$?
   g What is the value of x when $y = 0$?

2 Each of parts **a** to **d** has a line and some points.
   The points lie on the line.
   Find the missing co-ordinates.
   a $y = x + 5$ (3, ...)   (0, ...)   (-1, ...)
   b $y = x - 3$ (5, ...)   (3, ...)   (1, ...)
   c $y = 2x + 4$ (2, ...)   (0, ...)   (-1, ...)
   d $y = 3x - 4$ (4, ...)   (1, ...)   (-2, ...)

3 Write down the equation of each red line.

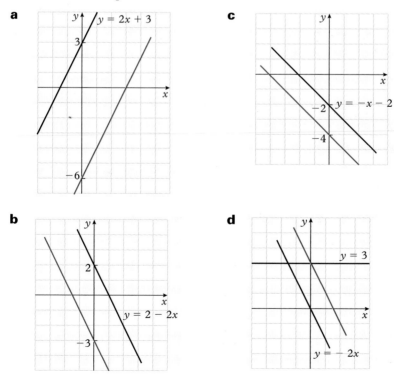

4 Solve these pairs of simultaneous equations.
You need to decide whether to add or subtract the equations.

a $6x + 2y = 20$
$4x + 2y = 14$

e $2x - 3y = 4$
$x + 3y = 11$

b $5x + 3y = 33$
$x + 3y = 21$

f $4x + 2y = 22$
$x + 2y = 10$

c $7x + 3y = 65$
$4x - 3y = 23$

g $7x + 3y = 13$
$4x + 3y = 10$

d $6x - y = 20$
$3x + y = 25$

h $2x - y = 4$
$x + y = 11$

5 Solve these pairs of simultaneous equations.

a $7a - b = 8$
$5a + 2b = 22$

c $5x + 2y = 63$
$4x - 3y = 32$

b $3x - 2y = 16$
$4x + 3y = 44$

d $7p - 3q = 23$
$5p + 4q = 41$

6 Solve these pairs of simultaneous equations.

a $3a = 21 - b$
$2a + b = 17$

c $17 = x + y$
$22 - 2y = x$

b $3x - 2y - 5 = 0$
$4x = 44 - 2y$

d $22 - r = 5 + s$
$3s = 2r + 1$

7 Find two points on each of these lines.
Do not draw the graph.

a $2x + 3y = 12$

c $3x + 4y = 24$

b $x + 3y = 9$

d $5x + 4y = 20$

8 The points $(1, 5)$ and $(2, 7)$ lie on the line $y = ax + b$

    a Substitute the values of x and y into the equation of the line.
$... = ...a + b$ and $... = ...a + b$
You now have two simultaneous equations.

    b Solve the two simultaneous equations to find the values of a and b.

9 The points $(1, 7)$ and $(2, 12)$ lie on the line $y = ax + b$

    a Substitute the values of x and y into the equation of the line to get two simultaneous equations.

    b Solve the two simultaneous equations to find the values of a and b.

10 The sum of two numbers is 12. The difference is 2.
Let the two numbers be x and y.
The sum of the two numbers is $x + y$
The difference of the two numbers is $x - y$
Write down two equations $\qquad x + y = ...$
$\qquad\qquad\qquad\qquad\qquad\qquad x - y = ...$
Find the two numbers by solving this pair of simultaneous equations.

11 The sum of two numbers is 22. The difference is 4.
Find the two numbers by solving a pair of simultaneous equations.

12 Find the region that is described by each of these sets of inequalities.

    a $y \geqslant -4 \qquad x \leqslant 3 \qquad y < x$

    b $x > 0 \qquad y \leqslant 2 \qquad y \geqslant x - 3$

    c $x > 1 \qquad y \leqslant 2 \qquad y \geqslant x$

    d $y \leqslant 5 \qquad y \geqslant x \qquad y \geqslant -x$

    e $y < 4 \qquad y \geqslant x + 1 \qquad y \geqslant 1 - x$

    f $y \geqslant x \qquad y \leqslant 2x \qquad y < 3$

1 Sam throws a ball into the air.
The distance travelled s is given by the formula
$$s = at - bt^2$$
When $t = 2$, $s = 20$ and when $t = 3$, $s = 15$
Find a and b.

2 Sally has science and maths homework to do.
She plans to spend less than 3 hours on maths.
The inequality $m < 3$ describes this.
m and s are the times spent on maths and science respectively.
a What do these inequalities describe?
   (1) $1 < s < 2$ (2) $m > s$
b Can m or s be negative?
c Draw a graph to show the region that contains all the possible values of m and s that satisfy all three inequalities.
   Make m the horizontal axis and s the vertical axis.

3 Find the set of inequalities that describe each of these regions.

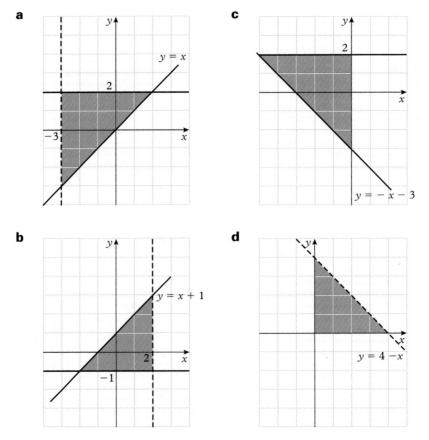

- **Gradient**

The **gradient** of a line tells you how steep the line is. Both these lines have a '3' to tell you how steep they are. They have **gradient** 3. The lines are parallel because they have the same gradient. The $+2$ and the -4 tell you where the lines cross the y axis.

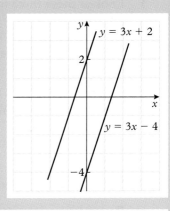

- **Simultaneous equations**

When you solve two equations at the same time you are are solving **simultaneous equations**.

Example

Solve this pair of simultaneous equations

$$5x + y = 20$$
$$2x + y = 11$$

Number the equations

$$(1)\ 5x + y = 20$$
$$(2)\ 2x + y = 11$$

Subtract to get rid of y
This finds x

$$3x \qquad = 9$$
$$x = 3$$

Use equation (1) to find y

Put $x = 3$ in equation (1)
$$5 \times 3 + y = 20$$
$$15 + y = 20$$
$$y = 5$$
The answer is $x = 3$, $y = 5$

Use equation (2) to check your answer
$$2x + y = 2 \times 3 + 5 = 6 + 5 = 11\ \checkmark$$

- **Inequalities**

You show the solutions to inequalities in 2 dimensions by shading regions.

Example

Find the region that is described by these inequalities.

$$y \leqslant 4 \qquad x > -1 \qquad y \geqslant x + 1$$

Shade the region for $y \leqslant 4$
Shade the region for $x > -1$
Shade the region for $y \geqslant x + 1$

The answer is the region R which is covered by all three shadings.

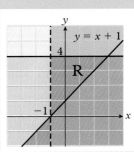

1 Look at these three lines.
$$y = 2x + 5 \qquad y = 5x - 2 \qquad y = 4x + 3$$
a Which line is the steepest?
b Where does the least steep line cross the y axis?

2 Each of these points lies on one of the lines.
Match the points with the lines.
$$(3, 7) \qquad (9, 5) \qquad (3, 2)$$
$$y = 2x - 4 \qquad y = x - 4 \qquad y = 3x - 2$$

3 Write down the equation of each red line.

a   **b**

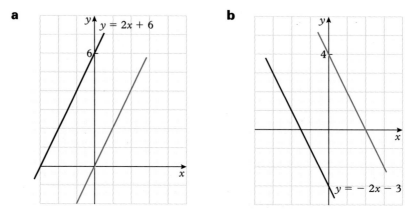

4 Solve these pairs of simultaneous equations.
a $2x + y = 11$ **b** $7x + 2y = 31$
    $5x - y = 17$ $3x + 5y = 34$

5 Find the region that is described by these three inequalities.
$$x \geqslant 2 \qquad y < 2 \qquad y \geqslant 3 - x$$

14 Graphs: on the button

You plot co-ordinates using x and y axes.

Computers can plot graphs using 3 dimensions.

This is a graph of $z = \sin xy$.

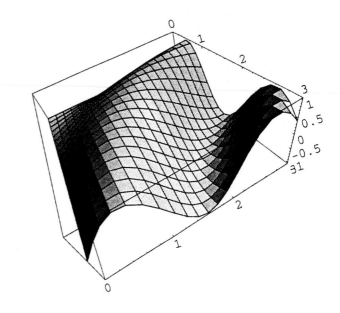

1   Graphs of calculator buttons

· ·

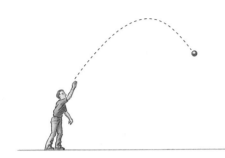

David is throwing a cricket ball.
The path of the ball makes a
curve in the air.
This type of curve is called a
parabola.

Exercise 14:1   *Graph of y = x^2*

1   **a**   Copy the axes on to
graph paper.
The x axis goes from -4 to 4.
The y axis goes from 0 to 16.
The scale is 1 cm to 1 unit on
both axes.

    b   Copy this table.
Fill it in. You can use x^2 on
your calculator.

x	0	1	2	3	4
$y = x^2$					

    c   Plot the points from your table.
    d   Join the points with a smooth
curve
Label the curve $y = x^2$

You have drawn part of a
parabola.

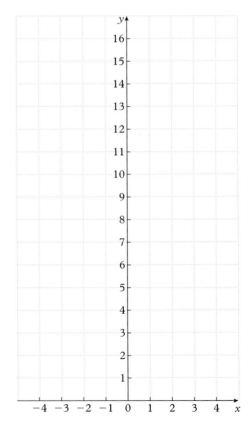

2 **a** Copy this table.

x	-4	-3	-2	-1	0
$y = x^2$					

b Fill in the table. You can use a calculator.

(1) Key in **4** **+/–** to get -4 on the calculator display.
Press **x^2**
Enter the answer in your table.

(2) Fill in the rest of the table in the same way.

c Plot the points from your table on the same diagram as question **1**.

d Join the points with a smooth curve.

· ·

Exercise 14:2 Graph of $y = \sqrt{x}$

1 **a** On graph paper, draw an x axis from 0 to 16 and a y axis from -4 to 4. Use a scale of 1 cm to 1 unit on both axes.

b Copy this table.
Fill it in. Use **$\sqrt{\ \ }$** on your calculator.
Give the values of y correct to 1 dp when you need to round.

x	0	1	2	4	6	8	10	12	14	16
$y = \sqrt{x}$										

c Plot the points from your table.
Join the points with a smooth curve.
Label your curve $y = \sqrt{x}$.

2 Calculators give positive square roots only.
Square roots can also have negative values.
Find $\sqrt{2}$ using your calculator.
Do not cancel the display but press **+/–**
This gives you the negative square root of 2.
Without cancelling the display press **x^2**
You should get back to 2.

3 a Copy this table.
Fill in the negative square roots.
Give the numbers correct to 1 dp when you need to round.

x	0	1	2	4	6	8	10	12	14	16
$y = -\sqrt{x}$			-1.4							

b Plot the points from your table on the same diagram as question **1**.
Join the points with a smooth curve.
c Compare your completed curve with the graph of $y = x^2$
Write down what you notice.

Exercise 14:3   Graph of $y = x^3$

1 a Copy these axes on to graph paper.
The x axis goes from -3 to 3 using a scale of 2 cm to 1 unit.
The y axis goes from -30 to 30 using a scale of 2 cm to 10 units.

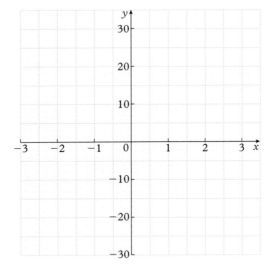

b Copy the table and fill it in.
You can work out x^3 as $x \times x \times x$ or you can use $\boxed{x^y}$ on your calculator.

x	-3	-2	-1	0	1	2	3
$y = x^3$							

c Plot the points from your table.
Join them with a smooth curve.
Label your graph $y = x^3$

Exercise 14:4 Graph of $y = \dfrac{1}{x}$

1 a Draw x and y axes from -5 to 5.
Use a scale of 1 cm to 1 unit on each axis.

b Copy this table.
Fill it in. You can use **1/x** on your calculator.
Give the values of y correct to 2 dp when you need to round.

x	0.2	0.3	0.4	0.5	1	2	3	4	5
$y = \dfrac{1}{x}$									

c Plot the points from your table.
Join them with a smooth curve.
Label the curve.

2 a Copy the table and fill it in.

x	-5	-4	-3	-2	-1	-0.5	-0.4	-0.3	-0.2
$y = \dfrac{1}{x}$									

b Plot the points from your table on the same diagram as question **1**.
Join the points with a smooth curve.

3 The curve is in two parts.
You need to decide what happens to the curve at 0.

a Cancel any calculator display so that your calculator shows 0.
Press **1/x**
What does your calculator show?

b $\dfrac{1}{x}$ means $1 \div x$
Work out $1 \div 0$ on your calculator.
What does your calculator show?

 c What do you think happens to the graph of $y = \dfrac{1}{x}$ when $x = 0$?

● **4 a** Write down a value of x between 0 and 0.2

Find $\dfrac{1}{x}$ for this value of x.

b Write down an even smaller positive value of x.

Find $\dfrac{1}{x}$ for this value.

c Continue finding $\dfrac{1}{x}$ for smaller and smaller values of x until you are

certain about what happens to the graph between 0 and 0.2

d Repeat parts **a** to **c** for values of x between -0.2 and 0

e Was your prediction in question **3** part **c** correct?

Different parabolas

Use graph paper for the graphs in this investigation.
Do not make your diagrams too small.

You do not need to use the same scale for x and y.
You could use 2 cm to one unit for x and 1 cm to 1 unit for y.

Remember: Changing the scale can change the shape of the curve.

1 Can you predict what the graph of $y = x^2 + 3$ will look like?
Draw a graph to see if you are right.

2 Investigate the shape of the graph of $y = x^2 + c$ where c is any number.

3 a Draw a graph of $y = x^2$
b Draw the graph of $y = 2x^2$. Use the same diagram.
c How does the 2 change the shape of the graph?
d Investigate the effect of different numbers in front of x^2.

2 Graph sketching

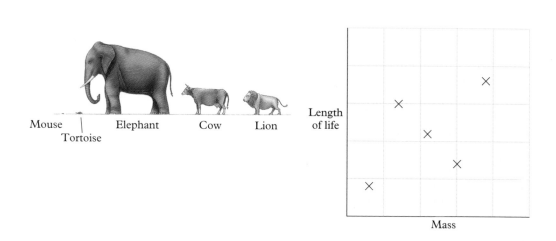

Mouse | Elephant   Cow   Lion
   Tortoise

Length of life

Mass

The graph shows the mass and the length of life of the 5 animals in the picture.

Can you decide which point represents each animal?

The graph does not have any scales on the axes. It is just a sketch graph. It is used to show a general pattern rather than to take accurate readings.

Exercise 14:5

1   Look at these pictures of cars.
Copy the axes shown and mark a point for each car.

Cost

Size of car

2 a Copy the axes shown.

b Six people took a science test and a maths test.
Here are some descriptions of how they did in the tests.
Daniel: Did well in both maths and science.
Kirsty: Did well in science but poorly in maths.
Richard: Was ill when he took the tests. He did badly in both subjects.
Anisha: Did very well in maths but only about average in science.
Nathan: Got an average mark in both tests.
Catherine: Did well in science but missed the maths test and so scored 0.

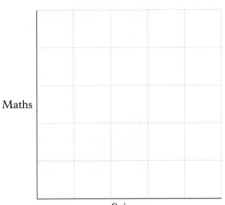

Mark a point on your graph to show each person's test results. Label each point.

3 Look at this graph. It shows the cost of some boxes of chocolates and the numbers of chocolates in each box.

a Which box is the most expensive?

b Which box is the best value for money?

c Which boxes cost the same?

d Which boxes have the same number of chocolate in them?

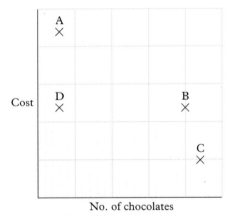

These two planes have different cruising speeds.
The smaller plane cruises at 230 mph.
The larger plane cruises at 350 mph.

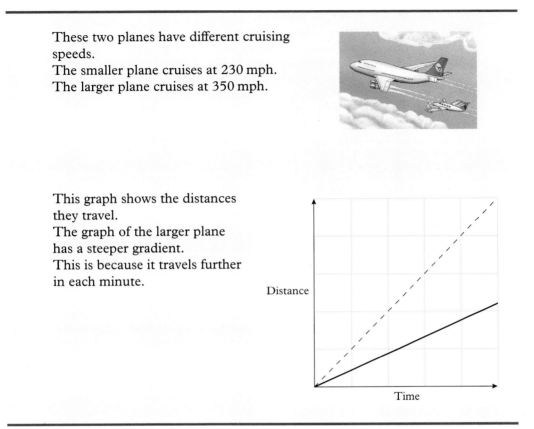

This graph shows the distances they travel.
The graph of the larger plane has a steeper gradient.
This is because it travels further in each minute.

Distance

Time

Exercise 14:6

For each question, sketch the graphs on the axes shown.
Think carefully about the gradients. Label each graph.

1 Car A has a cruising speed of 60 mph. Car B has a cruising speed of 50 mph.

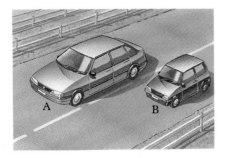

Distance

Time

2 These two fire hoses deliver water at different rates. The smaller hose delivers water at 1500 litres per second. The larger hose delivers water at 2300 litres per second.

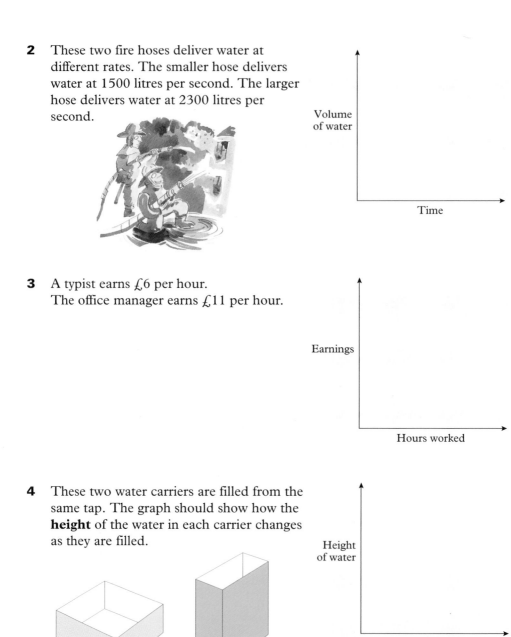

Volume of water

Time

3 A typist earns £6 per hour.
The office manager earns £11 per hour.

Earnings

Hours worked

4 These two water carriers are filled from the same tap. The graph should show how the **height** of the water in each carrier changes as they are filled.

Height of water

Time

5 The two water carriers in question **4** have the same volume. Which one will fill up first?

Gradient of a curve	The gradient of a straight line is the same all the way along. The **gradient of a curve** changes.

Example

A car accelerates rapidly up to 30 mph. It continues to accelerate up to 60 mph but its speed increases less rapidly. When it reaches 60 mph it travels at a steady speed. Sketch a graph to show the car's acceleration.

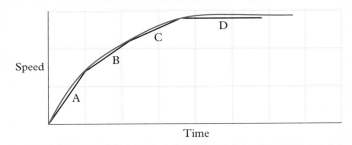

1 Use straight lines to find the shape of the graph. Line A is steep because the speed is changing rapidly. Lines B and C are less steep. The car's speed is still increasing but less rapidly. Line D is horizontal because the car reaches its steady speed.

2 Draw a curve to fit the pattern of the lines.

Sketch graph	A **sketch graph** can show a pattern or trend. You can't make accurate readings from it.

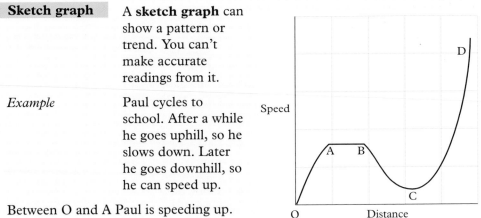

Example

Paul cycles to school. After a while he goes uphill, so he slows down. Later he goes downhill, so he can speed up.

Between O and A Paul is speeding up. The graph goes up.
From A to B he is going at a steady speed. The graph is horizontal. From B to C he is slowing down because he is going uphill. From C to D he is speeding up because he is going downhill. His speed at D is greater than it was at A.

Exercise 14:7

For each of these questions:
a Copy the axes.
b Draw a sketch of the graph from the description.
c Mark letters at important points and describe each section of the graph, like the second example on the previous page.

1 Howard cycles to school. The first part of his journey is along a level road. He then speeds up as he goes downhill. He slows down as he arrives at the school gate.

Speed | Distance

2 Lindsey goes to the local shops.
She jogs for the first part of her journey along a level road.
The next part is uphill and she walks quite slowly.
She stops at the shops at the top of the hill.

Speed | Distance

3 Katy made a jelly with boiling water.
She put it in the fridge to cool.
At first, it cooled very quickly and then it gradually settled down to a steady temperature.

Temp. | Time

4 Andy is running a computer games stall at the school Christmas Fayre. He thinks that if the games are too cheap he will not make much money. If they are too expensive, very few people will want to play.

Profit | Cost of 1 game

5 An apple grower is harvesting her crop. If she has too few people helping it will take a long time. If there are too many they will get in one another's way.

Time taken | Number of helpers

Anna is very keen on go-karting. She likes to visit different tracks to practise. It takes time for her to get used to a track and find out how quickly she can go on different parts of it.

Here is the diagram of Anna's home track.
She can go faster on the straights but she slows down for the corners.
The graph shows Anna's speed as she goes around the track.

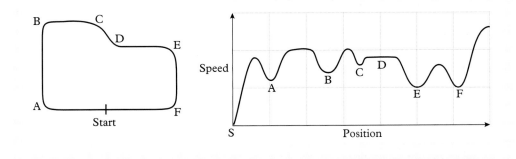

Exercise 14:8

For each of the go-kart tracks in questions **1** to **3** sketch a graph of the speed of the kart as it moves around the track. Use the letters to help you.

1

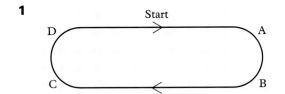

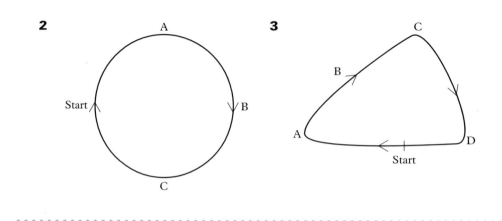

2

3

. . . .

Exercise 14:9   Designing a water tank

A water company is looking at new designs for water storage tanks.
The tanks need to hold 10 000 litres of water.
Here are some of the designs that have been suggested.

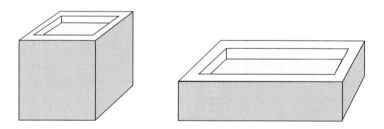

1   For each of the designs sketch a graph of the **height** of the water as
the tank is filled up using a hose. Think about the **rate** at which the
height would change.

2   Look at this design.
  a   Describe the main difference between
     this and the other three designs.
  b   What effect will this have on the graph
     of the height as it is filled up?
  c   Sketch the graph.

If the water tank has sloping sides, then the gradient of the graph will not be constant.
This means it will be a curve, not a straight line.

Example

Sketch a graph to show the height of the water in this tank as it is filled.

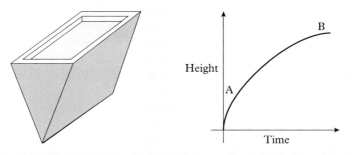

At A, the graph rises steeply because the tank is very narrow.
At B, the graph rises less steeply because the tank is much wider.

3 Sketch a graph to go with each of these tanks.

a   **b**   **c**

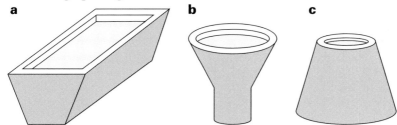

4 Here is a graph for another water tank.

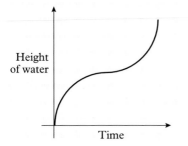

Draw a sketch of the tank by looking at the graph.

5 Design some water tanks of your own and sketch the graphs that go with them.

1   The table shows the values of x and y for the equation $y = x^2 + x - 2$

x	-4	-3	-2	-1	0	1	2	3
y		4						10

a Copy the table and fill it in.

b Draw an x axis from -4 to 3 and a y axis from -3 to 11.
Use a scale of 1 cm to 1 unit on both axes.
Plot the points from your table.
Join the points with a smooth curve.
Label your curve with its equation.

c Write down the co-ordinates of the point where the graph cuts the y axis.

d Write down the co-ordinates of the points where the graph cuts the x axis.

e Your curve should have line symmetry.
Write down the equation of the line of symmetry.

2   **a** Draw x and y axes from -12 to 12.
Use a scale of 1 cm to 2 units on both axes.

b (1) Complete this table for $y = \dfrac{12}{x}$

x	1	2	3	4	5	6	8	10	12
y									

(2) Plot the points from your table.
Join the points with a smooth curve.
Label your curve with its equation.

c (1) Complete this table for $y = \dfrac{12}{x}$

x	-12	-10	-8	-6	-5	-4	-3	-2	-1
y									

(2) Plot the points from your table.
Join the points with a smooth curve.

d Describe what happens to the graph between $x = -1$ and $x = 1$.

3 Ned is doing a parachute jump for charity.
As he jumps out of the plane, his speed increases rapidly.
He slows down suddenly as his parachute opens and then he floats to
the ground at a steady speed.

   a Which of these three graphs shows Ned's jump?
Explain your choice.

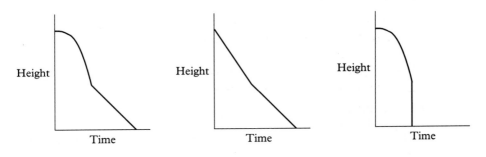

   b Explain why the other two graphs are wrong.
   c Sketch a graph of Ned's speed as he falls to the ground.

4 A roller coaster goes along this track.

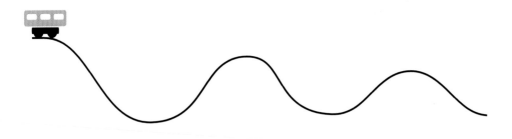

   a Sketch a graph of the height of the roller coaster above the ground
as it moves along the track. What do you notice?
   b Sketch a graph of the **speed** of the roller coaster as it moves along
the track.
   c Explain the connection between your two graphs.

1 Don has drawn this rectangle.
The formula for the area of Don's
rectangle is $x(6 - x)$.
Don wants to find the value of x
that makes the area of his rectangle
7 cm².
He needs to solve the equation
$x(6 - x) = 7$

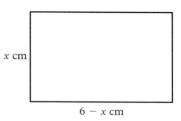

x cm

$6 - x$ cm

You can draw a graph to help you solve Don's equation.
a Draw an x axis from 0 to 6 and a y axis from 0 to 10.
Use a scale of 1 cm to 1 unit on both axes.
Use the axes to draw a graph of $y = x(6 - x)$ from $x = 0$ to $x = 6$.
b Draw the line $y = 7$ using the same axes.
Estimate the values of x where $y = 7$ crosses the graph $y = x(6 - x)$.
c There are two values of x that make the area of the rectangle 7 cm².
Explain why they are both correct.
d Use trial and improvement to find these values of x correct to 2 dp.

2 During the summer of 1995, the south west of England had swarms of
ladybirds. This was because there were lots of greenfly around.
Ladybirds live on greenfly!
This graph shows the number of greenfly during the summer.

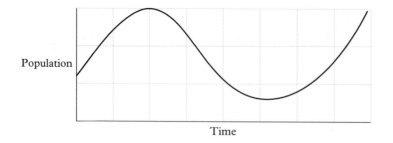

Population

Time

a Make a copy of this sketch graph. It does not have to be exact.
b Sketch the ladybird population on your copy of the graph.
Think carefully about how long it will take for the ladybird
population to increase.

- Examples of different shapes of graphs

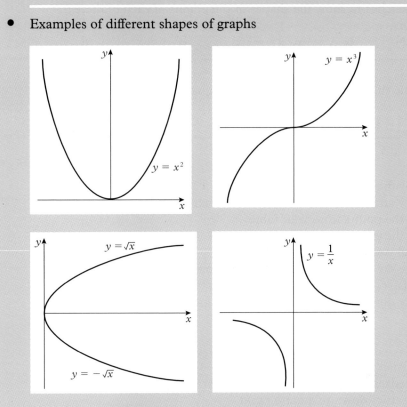

- **Sketch graph**   A **sketch graph** can show a pattern or trend.
You can't make accurate readings from it.

Example   Paul cycles to school. After a while he goes uphill, so he slows
down. Later he goes downhill, so he can speed up.

Between O and A Paul
is speeding up.
From A to B he is going
at a steady speed.
From B to C he is
slowing down because
he is going uphill.
From C to D he is
speeding up because he
is going downhill.

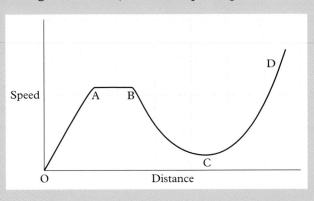

1 Draw sketches to show the shapes of the graphs of each of these equations.

  a $y = x^2$ **b** $y = \sqrt{x}$, $y = -\sqrt{x}$ **c** $y = x^3$ **d** $y = \dfrac{1}{x}$

2 The table shows the values of x for the equation $y = x^2 + x$

x	−4	−3	−2	−1	0	1	2	3
y		6				2		

  a Copy the table and fill it in.
  b Draw an x axis from −4 to 3 and a y axis from −2 to 12.
     Plot the points from your table.
     Join the points with a smooth curve.
     Label your curve with its equation.
  c Write down the co-ordinates of the points where the graph cuts the x axis.
  d Write down the co-ordinate of the point where the graph cuts the y axis.

3 When Brian goes cycling, his heart rate speeds up quickly at first.
It then settles down to a quite rapid but steady speed.
When he stops, it slowly returns to normal.

  a Copy these axes.
  b Sketch the graph of Brian's heart rate.

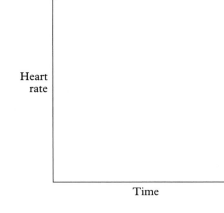

Heart rate

Time

4 This is the plan of a go-kart track. Sketch the graph of the speed of a go-kart as it goes around the track.

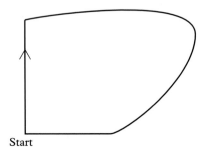

Start

15 Transformations

Ethnic designs often use the idea of transformations.

Can you see translations, rotations and reflections in this design?

1 Transformations

Sian has moved the furniture in her bedroom. She has turned the bed through 90° so that it fits into the corner.
The desk has been moved along the wall.
The poster on the wall is an enlargement of a photograph Sian took of her dog.
The mirror shows Sian's reflection.

In mathematics any movement of a shape is called a transformation.
There are four main types of transformations.
Reflection   Translation   Rotation   Enlargement

◄◄REPLAY►

Object	The shape you start with is called the **object**.
Image	The transformed shape is called the **image**.
Congruent	When shapes are identical they are called **congruent**.

This shape has been reflected.

The object and image are congruent.

Line of reflection

Line of reflection	The mirror line is called a **line of reflection**.

Exercise 15:1

1 These shapes have been reflected.
Write down the equation of the line of reflection for each diagram.

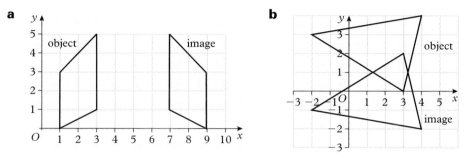

Sometimes it can be hard to find the line of reflection.

To find the line of reflection

Shape A is reflected to get shape B.
The line of reflection is halfway
between A and B.

Choose a point P on shape A.
Find the corresponding point on
shape B. Mark this P'.

Join P and P' with a line.
Construct the perpendicular
bisector of the line PP'.

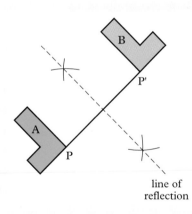

The bisector is halfway between
A and B.
The bisector is the line of reflection.

2 Ask your teacher for a copy of worksheet 15:1.

◄◄REPLAY►

| Translation | A **translation** is a movement in a straight line. |

Example

This object and image show a translation.
Describe the translation.

Mark a point P on the object.
Mark the corresponding point on the image. Label it P'.

P has moved 3 units to the right and 1 unit up.
This means that the whole shape has moved 3 units to the right and 1 unit up.
The object and image are congruent.

Exercise 15:2

For questions **1** to **8** describe the translation that is needed to move the object to the image.

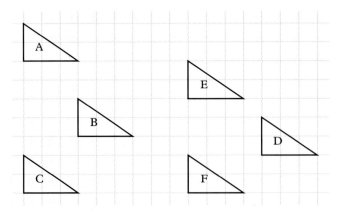

	Object	Image			Object	Image
1	A	B		**5**	B	C
2	C	D		**6**	E	F
3	B	E		**7**	F	C
4	E	C		**8**	A	C

Rotation	A **rotation** turns a shape through an angle about a fixed point.
Centre of rotation	The fixed point is called the **centre of rotation**.

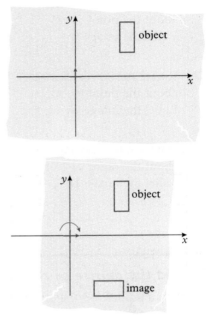

Nina is going to rotate this rectangle through 90° clockwise.
The origin is the centre of rotation.
Nina puts some tracing paper on top of the diagram. She traces the rectangle and marks a cross at the centre of rotation.

Nina uses the cross to help her to rotate the tracing paper through 90° clockwise.
She marks the new position of the rectangle on her axes. She labels this the image.

The object and image are congruent.

Exercise 15:3

1 Copy the axes and the shape S on to squared paper.

2 Rotate S 90° clockwise about the origin.
Use tracing paper to help you.
Label the image A.

3 Rotate S 180° anticlockwise about the origin.
Label the image B.

4 Rotate S 90° anticlockwise about the origin.
Label the image C.

5 Copy the axes and the shape A on to squared paper.

6 Rotate A anticlockwise about the point $(1, 2)$. Label the image B.

7 Rotate A $180°$ clockwise about the point $(1, 2)$. Label the image C.

8 Rotate A $90°$ clockwise about the point $(-1, 4)$. Label the image D.

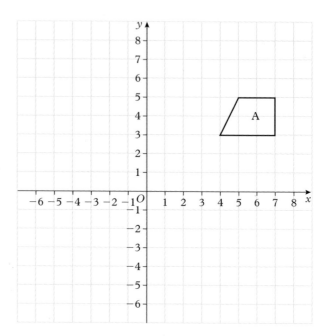

To find the centre of rotation

1 Mark a point P on the object. Join it to the corresponding point P′ on the image. Draw the perpendicular bisector of this line.

2 Do the same for a second pair of points Q and Q′. The point where the two bisectors cross is the centre of rotation, C. The centre of rotation is $(-1, 2)$. Join CP and CP′. The angle between CP and CP′ is $90°$. The angle of rotation is $90°$ clockwise.

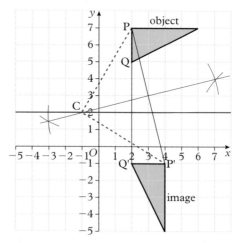

Exercise 15:4

Questions **1** to **6** show rotations.
For each question:
a Copy the diagram.
b Find the centre of rotation and write down its co-ordinates.
c Write down the angle of rotation, giving its direction.

1

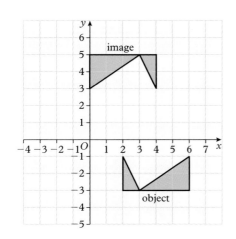

4

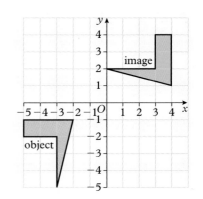

2

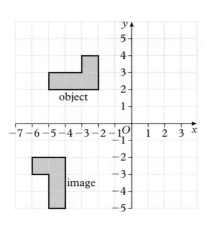

5

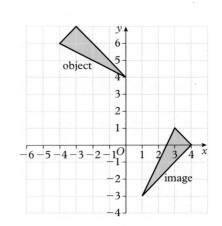

3

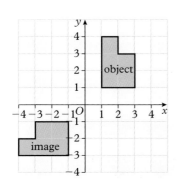

6

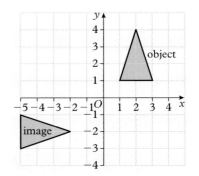

2 Combined transformations

Richard is putting up a display in the maths room to show some patterns made by transforming shapes.
He has made these patterns from a single shape.

This is the shape that Richard used.

Richard made this pattern by translating the shape.

He made this pattern by reflecting the shape in a vertical line of reflection.

Exercise 15:5

1 What transformation would you use with this shape to make each of these patterns?

a

b

2 Tara has made a puzzle for her brother Chris.
The puzzle has four cards placed like this.

Chris can only move the cards in three ways:

(1) He can reflect a card vertically (the line of reflection is vertical).

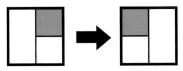

(2) He can reflect a card horizontally (the line of reflection is horizontal).

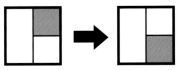

(3) He can rotate a card through an angle.

Chris has to transform the cards in the first pattern to make the second pattern.

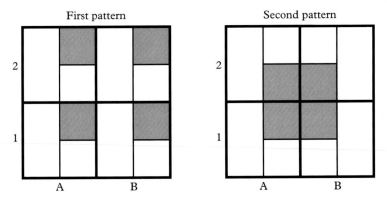

a Copy this table. It shows the transformations needed for cards A1 and B1 to make the second pattern.
b Fill in the transformations needed for A2 and B2.

Card	Transformation
A1	none – card is in correct position
B1	reflect vertically
A2	
B2	

Combined transformations

Shape A is reflected in the *x* axis to give the image B.

Shape B is rotated 180° clockwise about the origin to give the image C.

You need to find a **single** transformation that moves shape A to shape C.

The transformation is a reflection in the *y* axis.

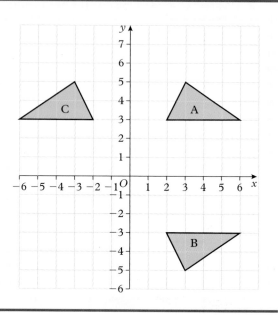

Exercise 15:6

1 **a** Copy this set of axes and shape P.
   b Reflect shape P in the *x* axis. Label the image Q.
   c Reflect shape Q in the *y* axis. Label the image R.
   d What single transformation will move shape P to shape R?

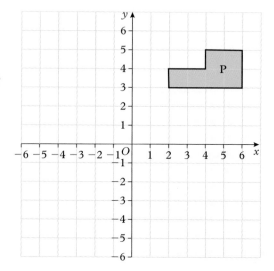

2 **a** Copy this set of axes and triangle A.
  b Reflect triangle A in the y axis. Label the image B.
  c Draw the line $y = x$. Reflect triangle B in the line $y = x$. Label the image C.
  d What single transformation will move triangle A to triangle C?

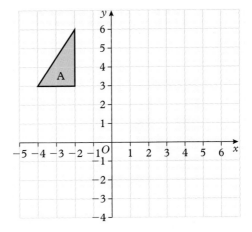

3 **a** Draw a set of axes with x and y from -6 to $+6$. Draw the triangle with vertices (2, 1), (3, 1) and (2, 4). Label this triangle A.
  b Reflect triangle A in the x axis to get triangle B. Then reflect triangle B in the y axis. Label the final image C.
  c Reflect triangle A in the y axis to get triangle D. Then reflect triangle D in the x axis to get triangle E. What do you notice about triangle E?

4 Draw another set of axes like those in question **3**. Draw and label triangle A. Use the triangle to check if the following statement is true:

'A reflection in the x axis followed by a rotation of 90° anticlockwise is the same as a rotation of 90° anticlockwise followed by a reflection in the x axis'.

3 Enlargement

Some things get blown up out of all proportion.

Enlargement   An **enlargement** changes the size of an object.
The change is the same in all directions.

Scale factor   The **scale factor** tells us how many times bigger the enlargement is.

Example   Enlarge the triangle by a scale factor of 2.

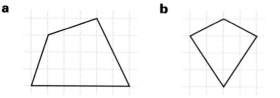

object   image

The enlargement is 2 times as long and 2 times as high as the object.
The object and image are **not** congruent.

Exercise 15:7

1 Copy these shapes on to squared paper.
Enlarge each shape by a scale factor of 2.

a   b   c

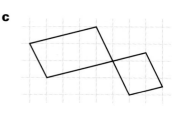

2 Enlarge the shapes in question **1** by a scale factor of 3.

15

The word **enlarge** normally means to make something **bigger**. In maths, enlargements can also make objects **smaller**! This happens when the scale factor is less than one.

Example

Enlarge the trapezium by a scale factor of $\frac{1}{2}$.

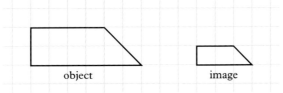

object   image

The enlargement is $\frac{1}{2}$ as long and $\frac{1}{2}$ as high as the object.

3 For each of these pairs of shapes, write down the scale factor of the enlargement that transforms:

a S into T   **b** T into S

Make any measurements that you need to help you.

(1)

(2)

4 Copy these shapes on to squared paper.
Enlarge each shape by a scale factor of $\frac{1}{2}$.

a

b

5 Enlarge the shapes in question **4** by a scale factor of $\frac{1}{3}$.

Centre of enlargement

An enlargement can be done from a **centre**.

Measure the distances of points on the object from the centre.
Multiply these distances by the scale factor.
This gives the distances of points on the image from the centre.

Example

Enlarge the black triangle using C as the centre of enlargement:

a by a scale factor of 2   **b** by a scale factor of $\frac{1}{2}$

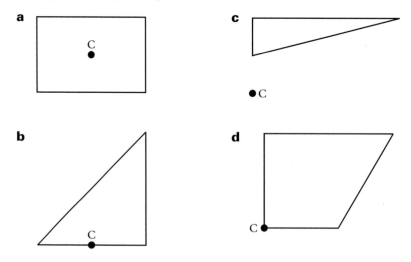

6 Copy these diagrams.
Use C as the centre of enlargement each time.
Enlarge each shape using a scale factor of 2.

a

c

b

d

7 Enlarge the shapes in question **6** using a scale factor of $\frac{1}{2}$.

Similar

If two objects have the same shape but different sizes they are **similar**. One is an enlargement of the other.
You can use the scale factor to find missing lengths.

Example

Rectangles P and Q are similar.
Find the missing lengths.

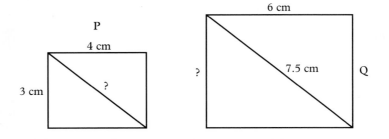

The ratio of the length of P to the length of Q is 4 : 6.
Convert the ratio to the form 1 : *n*.

$$4 : 6 = \frac{4}{4} : \frac{6}{4} = 1 : 1.5$$

n is the scale factor of the enlargement.
Here the scale factor is 1.5

The width of rectangle P is 3 cm.
The width of rectangle Q is $3 \times 1.5 = 4.5$ cm

The length of the diagonal of Q is 7.5 cm.
The length of the diagonal of P is $7.5 \div 1.5 = 5$ cm

Exercise 15:8

Each question in this exercise has a pair of similar shapes.
  a Write down the ratio of the red sides, smallest side first.
  b Convert the ratio to the form 1 : *n*.
  c Write down the value of the scale factor, *n*.
  d Use the scale factor to find the sides marked with letters.

1

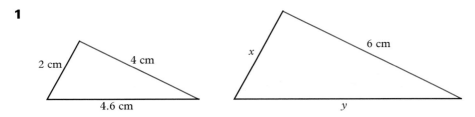

2

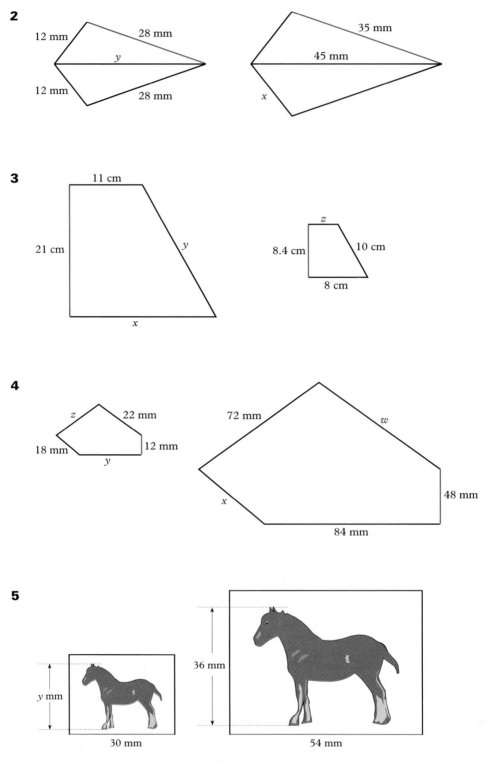

3

4

5

4 Similar triangles

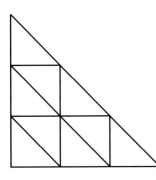

There are 13 similar triangles in this diagram.
Can you find them?

Similar triangles

Two triangles are **similar** if they have the same angles.

Example

Triangle P is similar to triangle Q.
Triangle P is the same shape as triangle Q.
This means that the angles of triangle P are equal to the angles of triangle Q.

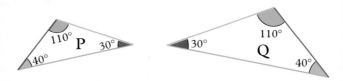

Corresponding angles and sides are marked in the same colours.

Exercise 15:9

For each pair of triangles in questions **1** to **3**:
a Sketch the diagram.
b Calculate any missing angles.
c Decide if the triangles are similar by checking the angles.
d Colour corresponding sides and angles if the triangles are similar in the same way as the example.

1

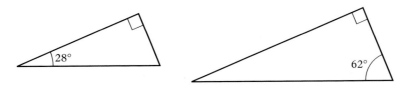

2 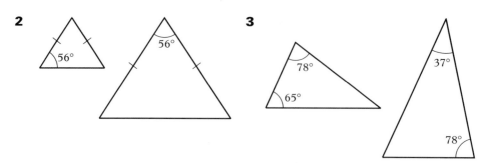 **3**

The angle properties of parallel lines can be used to find the angles of similar triangles.

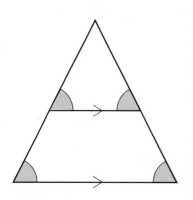

The colours show pairs of **corresponding angles** ('F' angles).

Alternate angles ('Z' angles) are shown in green and blue.
Opposite angles are red.

In questions **4** and **5**:
a Sketch the diagram.

b Colour pairs of equal angles.

4 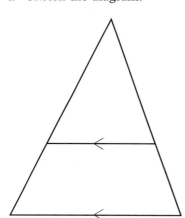 **5**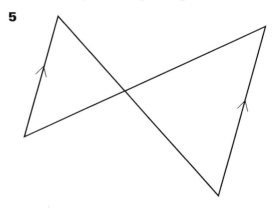

Look at the two triangles.

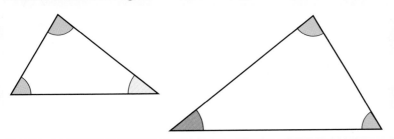

There are **two pairs** of **equal angles** (green and blue).
This means that the **third pair** of angles must be **equal** (the yellow angle
and the red angle).

In questions **6** and **7**:
a Sketch the diagram.

b Use three colours to show
three pairs of equal angles.

6 Angle EAD = angle EBC

7 Angle ACB = angle AFE

8 a Sketch the diagram.
b Use three colours to show three
pairs of equal angles.
c Use the same three colours to
mark corresponding sides.
d Find the scale factor of the
enlargement that transforms
triangle CDE into triangle ABC.
e Find the lengths of AC and DC.

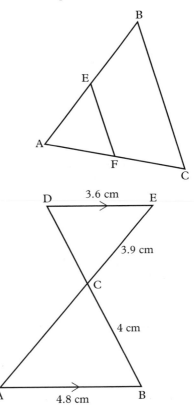

9 **a** Sketch the diagram.
   b Use two colours to show two pairs of
      equal angles.
      Use a third colour for angle P.
   c Use the same three colours to mark
      corresponding sides.
   d Use sides PS and PQ to work out
      the scale factor for the similar triangles.
   e Find the length of QR.

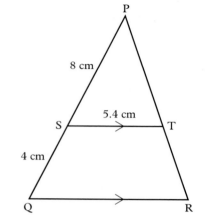

10 Look at the diagram.
   The two triangles AED and BEC
   are similar.
   Angle EDA is equal to angle ECB.
   AD = 45 mm, BC = 30 mm and
   DE = 39 mm
   Calculate the length of CE.

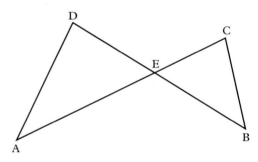

Exercise 15:10

1 Jason has a pair of steps he uses when
   he is decorating. The diagram shows
   the steps after Jason has opened them.
   There is a rope connecting the sides
   of the steps. The rope stops the steps
   from opening too far.
   Calculate the length of the rope.

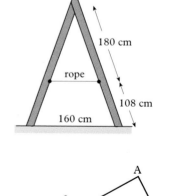

2 The Timber Supply Company
   make wooden frames for roofs
   of new houses. The diagram
   shows one of their designs.
   In this design
   angle ABC = angle FGC and
   angle ACB = angle EHB.
   a Calculate the length of GC.
   b Calculate the length of FG.
      Give your answer correct to 2 dp.

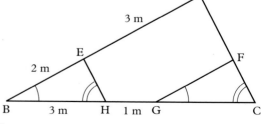

1   Copy the axes and shape P
on to squared paper.

   a Rotate shape P 90°
   anticlockwise about the
   point (2, 1).
   Label the image A.

   b Rotate shape P 180°
   anticlockwise about the
   point (2, 1).
   Label the image B.

   c Rotate shape P 90°
   clockwise about the point
   (2, 1).
   Label the image C.

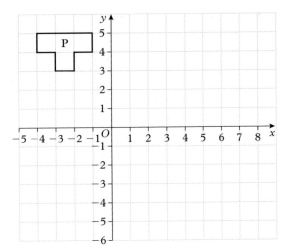

2   **a** Copy this set of axes and
   shape A.

   b Rotate shape A 90°
   anticlockwise with the
   origin as the centre of
   rotation.
   Label the image B.

   c Rotate shape B 180°
   clockwise with the origin
   as the centre of rotation.
   Label the image C.

   d What single
   transformation will move
   shape A to shape C?

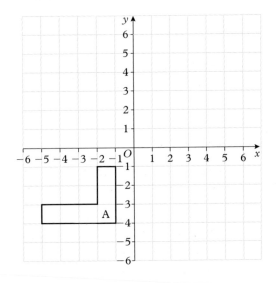

3   Nathan has made some different sized photocopies of the same house
shape.

   a Calculate the
   lengths marked
   with letters.

   b Write down the
   size of the angle
   marked $t°$.

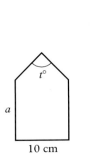 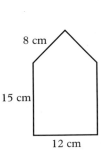 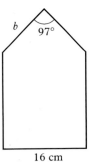

325

4 Draw an *x* axis from −10 to 10 and a *y* axis from −6 to 10 on to squared paper.
  a Plot the points A (0, 4) B (4, 2) C (4, −2) and D (−4, −2). Join the points to get quadrilateral ABCD.
  b Use the origin as the centre to draw the enlargement of ABCD, scale factor 2.
  c Use the origin as the centre to draw the enlargement of ABCD, scale factor $\frac{1}{2}$.

5 Calculate the length DE in each of these diagrams.

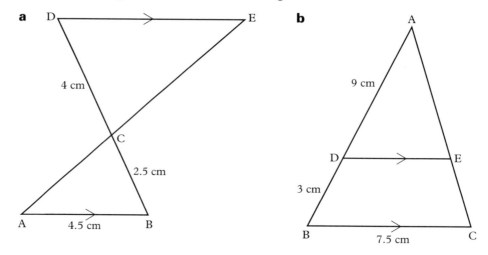

6 The diagram shows the side of Sapna's camping stool. Find the distance between the bottoms of the legs of the stool marked in the diagram.

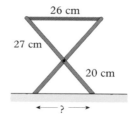

1 a Copy this set of axes and shape A.
  b Reflect shape A in the *x* axis. Label the image B.
  c Rotate shape B 90° clockwise with the origin as the centre of rotation. Label the image C.
  d What single transformation will move shape A to shape C?

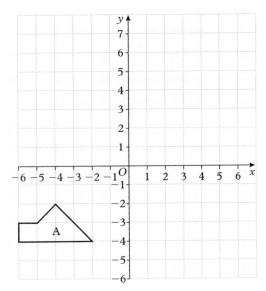

2 a Copy this set of axes and triangle A.
  b Reflect triangle A in the line $y = x$ Label the image B.
  c Reflect triangle B in the line $y = -x$ Label the image C.
  d What single transformation will move triangle A to triangle C?

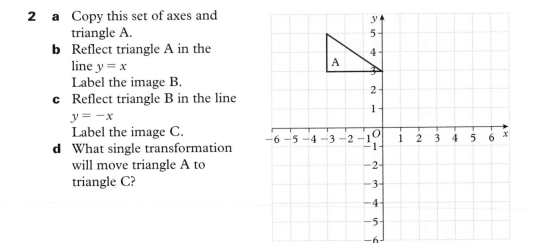

3 WXYZ is a rectangle with vertices W (2, 3), X (6, 3), Y (6, 5) and Z (2, 5).
Draw a set of axes on squared paper.
Draw and label rectangle WXYZ.
Plot the point C, (0, 1).
Using C as the centre of enlargement, enlarge WXYZ by a scale factor of $\frac{1}{2}$.

4 The diagram shows a design for a school badge.

a The badge has to fit on to a rectangular blazer pocket. The pocket is 15 cm high and 13 cm wide.
What scale factor will enlarge the badge so that it just fits the pocket?

b The badge is to be on bookmarks which are sold for charity. The bookmarks are 14 cm long and 2.5 cm wide.
What scale factor will enlarge the badge so that it just fits the bookmark?

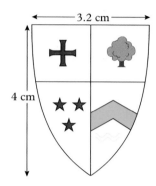

3.2 cm

4 cm

5 Two sheets of A4 paper placed side by side are the same size as a sheet of A3 paper.
Use the fact that A4 paper is 210 mm by 297 mm to work out the scale factor that enlarges A4 into A3.
Write down all the figures on your calculator display for the scale factor.
Do not cancel the display.
Press x^2 .

Round the answer to 3 sf.
Explain your answer.

6 The diagram shows a cross section of a swimming pool.
The pool is being emptied.
Work out the depth of water left in the pool.

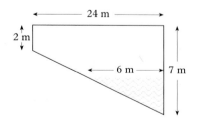

24 m

2 m

6 m   7 m

• **Translation**	A **translation** is a movement in a straight line.
• **Rotation**	A **rotation** turns a shape through an angle about a fixed point.
Centre of rotation	The fixed point is called the **centre of rotation**.
• **Congruent**	When shapes are identical they are called **congruent**. For reflection, translation and rotation the object and the image are always congruent. For enlargement they are not congruent.
• **Enlargement**	An **enlargement** changes the size of an object. The change is the same in all directions. The object and image are not congruent.
Scale factor	The scale factor tells us how many times bigger the enlargement is.
• **Similar triangles**	Two triangles are **similar** if they have the same angles.
Similar	If two objects have the same shape but different sizes they are **similar**. One is an enlargement of the other. There is a fixed scale factor or ratio between the sides. You can use the scale factor to find missing lengths.
Example	Rectangles P and Q are similar. Find the missing lengths.

The ratio of the length of P to the length of Q is $4 : 6$.

Convert the ratio to the form $1 : n$. $4 : 6 = \frac{4}{4} : \frac{6}{4} = 1 : 1.5$

n is the scale factor of the enlargement.
Here the scale factor is 1.5

The width of rectangle P is 3 cm.
The width of rectangle Q is $3 \times 1.5 = 4.5$ cm
The length of the diagonal of Q is 7.5 cm.
The length of the diagonal of P is $7.5 \div 1.5 = 5$ cm

1   **a**   Copy this set of axes
and the shape P.
    b   Reflect P in the *x* axis.
Label the new shape Q.
    c   Reflect P in the *y* axis.
Label the new shape R.
    d   Translate R 8 units down.
Label the new shape S.
    e   The single transformation
that will move Q to S
is a rotation.
Find the co-ordinates of
the centre of rotation.
Find also the angle
of the rotation.
    f   What single transformation
will move Q to R?

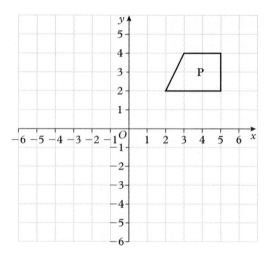

2   **a**   Draw this rectangle to scale.
    b   Use C as the centre of enlargement.
Enlarge the shape by a scale factor of:
     (1) 2   (2) $\frac{1}{2}$

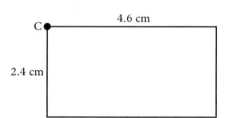

3   For each of these diagrams:
    a   Sketch the diagram and mark equal angles.
    b   Mark corresponding sides.
    c   Find the lengths marked with letters.

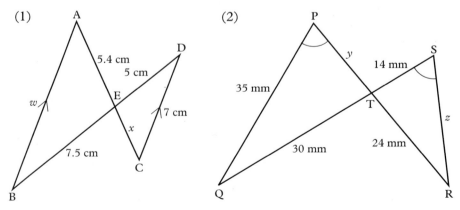

16 Get in shape

QUESTIONS

EXTENSION

SUMMARY

TEST YOURSELF

You can make a Möbius strip with a piece of card or paper.

Make a half twist (180°) before sticking the two ends together.

What happens if you draw a line down the centre of the strip and cut along it?

1 Polygons

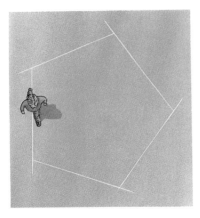

Look at this large pentagon marked out in a school yard.
Peter walks along the first side.

To walk along the second side he turns through the angle marked in red.

To walk along the third side he turns through the angle marked in blue.

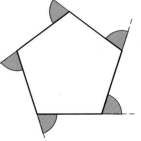

When Peter has walked all the way around the shape he is back facing in the same direction that he started.
He has turned through 360°.
The exterior angles of a pentagon must total 360°.

Exterior angles	The **exterior angles** of a shape are the angles outside the shape.

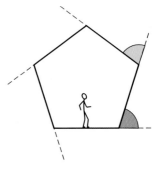

The sum of the exterior angles of any polygon is 360°.

Example

Work out the size of the exterior angle of a regular pentagon.

The sum of the angles is 360°.
If the pentagon is regular all its exterior angles will be equal.

$$\text{Exterior angle} = \frac{360°}{5} = 72°$$

Exercise 16:1

1   Work out the exterior angle of a regular octagon.

2   Work out the exterior angle of a regular nonagon (nine sides).

3   Work out the exterior angle of a regular decagon (10 sides).

4   Work out the exterior angle of a regular heptagon (seven sides).
    Round your answer correct to 1 dp.

5   Work out the exterior angles for regular polygons with the following
    numbers of sides.
    a 12   **b** 15   **c** 11   **d** 18

6   Four of the exterior angles of a pentagon are 56°, 47°, 103° and 67°.
    What is the size of the fifth angle?

7   The exterior angle of a regular polygon is 36°.
    How many sides has it got?

8   Explain why there is no regular polygon with an exterior angle of 35°.

9   This irregular pentagon is symmetrical.
    Work out the sizes of the exterior angles.

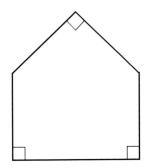

10 Two identical irregular quadrilaterals with
    the interior angles shown are put together
    to form a hexagon.
    a Calculate the exterior angles of the
          hexagon.
    b Check that the sum of the exterior
          angles is 360°.

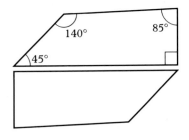

Exercise 16:2

1 In this question you are going to draw a regular pentagon.
You will need a protractor or angle measurer.

   a Draw a horizontal line 10 cm long, with 5 cm solid and 5 cm dotted.
Leave some space above it.

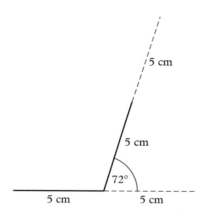

   b Put your protractor on the end of the solid line.
Measure 72° from the dotted line.

   c Draw another line 10 cm long.
Draw 5 cm solid and 5 cm dotted.
You should now have a diagram like the one shown.

   d Repeat **b** and **c** on your new line.

   e Keep going until you complete the pentagon.

2 Use the same method to draw each of these:
   a a regular hexagon,
   b a regular octagon.

Interior angles   The angles inside a polygon are called **interior angles**.
They always make a straight line with the exterior angle.

This means that exterior angle + interior angle = 180°.

interior angle = 180° − exterior angle

Example   Work out the interior angle of
a regular hexagon.

$$\text{Exterior angle} = \frac{360°}{6} = 60°$$

$$\text{Interior angle} = 180° - 60°$$
$$= 120°$$

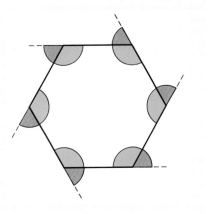

Exercise 16:3

1   Work out the interior angle of a regular pentagon.

2   Work out the interior angle of a regular octagon.

3   Copy this table. Fill it in.

Number of sides	Name of polygon	Exterior angle	Interior angle
3	equilateral triangle		
4	square		
5	regular pentagon		
6	regular ...		
7	regular ...		
8	regular ...		
9	regular ...		
10	regular ...		

Tessellation   A **tessellation** is a pattern made by repeating the same shape
over and over again.

There are no gaps in a tessellation.

4 Tessellations

You will need a polygon stencil or some templates of regular polygons.
Squares tessellate. They fit together without any gaps.

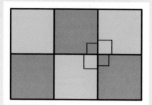

The angles around a
point total 360°.

If a regular polygon tessellates by itself it is called a **regular tessellation**.

If two regular polygons tessellate together this is called a **semi-regular tessellation**.

a Find as many regular tessellations as you can.
Mark on the interior angles and show that the total around a point
is 360°.

b There are 8 different semi-regular tessellations. Find them all.
Mark on the interior angles to show that the total around a point is 360°.

5 Explain why regular pentagons do not tessellate.
Use angles to help you.

6 **a** 'All triangles tessellate.'

Is this true or false?
Draw some triangle tessellations.
Measure and mark the angles.

b 'All quadrilaterals tessellate.'

Is this true or false?
Draw some quadrilateral tessellations.
Measure and mark the angles.

Sum of the interior angles

You can find the **sum of the interior angles** of any polygon.

(1) Draw the polygon.

(2) Join one vertex to all the others. All the interior angles are now inside one of the triangles.

(3) Count the number of triangles. Multiply by 180° to find the total.

Example

Find the sum of the interior angles of an octagon.

The octagon splits into 6 triangles.

Total = 6 × 180° = 1080°

If it is a regular octagon, each interior angle $= \dfrac{1080°}{8}$

$= 135°$

Exercise 16:4

For each of the polygons in questions **1**, **2** and **3**:
a Draw a sketch.
b Split the polygon into triangles.
c Work out the sum of the interior angles.
d Work out the size of each interior angle if the polygon is regular.
e Check your answers to **d** with your table from Exercise 16:3.

1 Pentagon

2 Hexagon

3 Heptagon

4 Calculate the missing angle in this irregular hexagon.

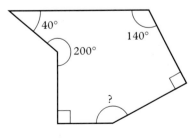

2 Symmetry

This building consists of two houses.
One house is the reflection of the other.

You cannot walk across the inside from one house to the other.
Where the houses join inside there is a solid vertical wall.
This wall is the plane of symmetry of the houses.

◀◀ REPLAY ▶

Line of symmetry	A **line of symmetry** divides a shape into two equal parts. Each part is a reflection of the other. A line of symmetry is also called an **axis** of symmetry.
Rotational symmetry	A shape has **rotational symmetry** if it fits on top of itself more than once as it makes a complete turn.
Order of rotational symmetry	The **order of rotational symmetry** is the number of times that the shape fits on top of itself. This must be 2 or more.

C marks the centre of rotation

A regular hexagon has 6 axes of symmetry.
It has rotational symmetry of order 6.

Exercise 16:5

For each of the polygons in questions **1**, **2** and **3**:
a Draw a sketch.
b Draw the axes of symmetry on your sketch.
c Mark the centre of symmetry on your sketch.
d Write down the order of rotational symmetry.

1 Equilateral triangle

2 Square

3 Pentagon

4 a Write down the number of axes of symmetry of a regular octagon.
   b Write down the order of rotational symmetry of a regular octagon.

Plane	A **plane** is a flat surface.
Plane of symmetry	A **plane of symmetry** divides a *solid* into two equal parts. Each part is a reflection of the other.

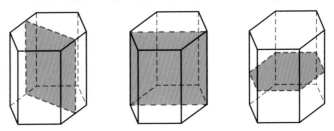

The sketches show three different planes of symmetry for a hexagonal prism.

---

5 The base of a prism is a regular hexagon.
   a How many vertical planes of symmetry does the prism have?
   b How many horizontal planes of symmetry does it have?
   c How many planes of symmetry does the prism have altogether?
   The red lines are called axes.
   d Write down the order of rotational symmetry of the prism about the axis:
   (1) AA′   (2) BB′

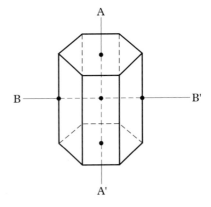

6 The bases of these prisms are regular polygons.
Write down the total number of planes of symmetry of each prism.

(1)   (2)   (3)

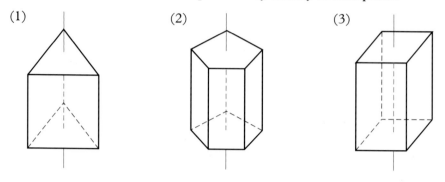

7 Each of these pyramids has a regular polygon for its base.
  a Write down the number of planes of symmetry for each pyramid.
  b Write down the order of rotational symmetry for each pyramid about its vertical axis.

(1)   (2)   (3)

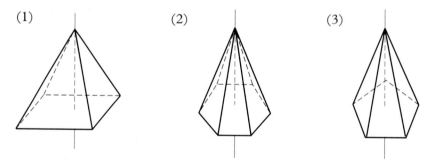

8 **a** Write down the number of planes of symmetry of each of these objects.
  b List the objects that have rotational symmetry.

3   Nets of solids

Katie has a chocolate bar.
The box for the bar is in the
shape of a triangular prism.

Katie wants to see how the box
was made. She has opened out
the card.

◄◄REPLAY►

Net

A **net** is a pattern of shapes on a piece of paper or card.
The shapes are arranged so that the net can be folded to make
a hollow solid. Notice that a net does not have flaps.

A cuboid has six faces but they are not all the same size.

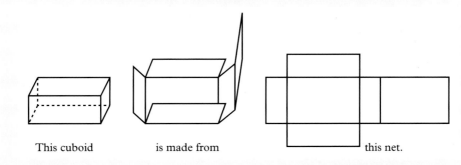

This cuboid   is made from   this net.

Exercise 16:6

1 Use a piece of squared paper to
draw an accurate net of this cuboid.
Do not cut the net out.
Keep your net to use in question **2**.

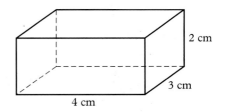

2 cm

3 cm

4 cm

To make a solid you need to add flaps to the net.

Half the edges of these nets have flaps. The flaps are on alternate edges.

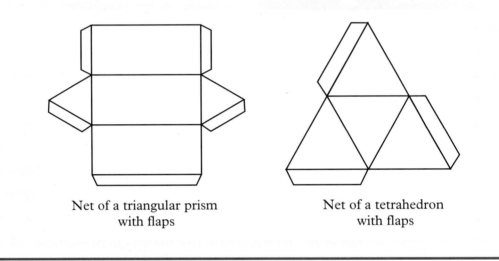

Net of a triangular prism
with flaps

Net of a tetrahedron
with flaps

2 **a** Add flaps to the net of the cuboid you drew in question **1**.
Put them on alternate edges.
Make the flaps just less than 1 cm wide.
If they are too narrow they will be difficult to stick.
b Cut out the net and fold it to make a cuboid.
Check that the flaps are on the correct edges.

3 **a** Use plain paper to draw an accurate net of each of these solids.
Use compasses to construct the triangles accurately.
b Add flaps to each of your nets.
Put them on alternate edges.
c Cut out each net and fold it to make a solid.
Check that the flaps are on the correct edges.

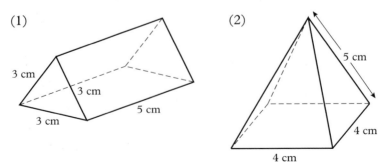

(1)

3 cm

3 cm

3 cm

5 cm

(2)

5 cm

4 cm

4 cm

4 The polygons at the ends of these prisms are coloured red.
The faces at the sides of the prisms are yellow.

   a The yellow faces are all the same shape.
     Write down the name of this shape.

   b Each solid has two red faces.
     Write down the number of yellow faces that each solid has.

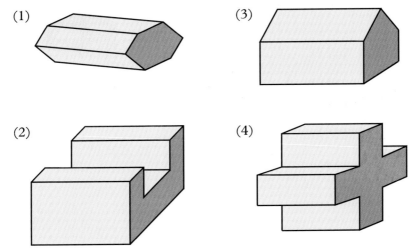

(1)   (3)

(2)   (4)

5 Choose one of the shapes in question **4** and sketch its net.

6 Write down the names of the solids that can be made from these nets.

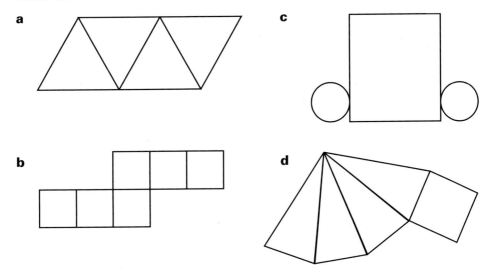

   a   **c**

   b   **d**

4 Building a shape sorter

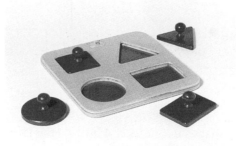

Look at these children's toys. They are shape sorters.

They are used to improve children's motor skills and co-ordination.

They also help to teach children about shape and colour.

It is very important that each shape will only fit through its own hole. If it fits through another hole, the child will not learn to match the shape of the piece with the shape of the hole.

Most shape sorters have at least six shapes in them.

The holes are often regular polygons or symmetrical shapes.

This makes it easier for the child as there are more ways that a shape will fit through the right hole.

Most of the shapes are prisms.
They have the same cross section along their whole length.

Building a shape sorter

Build a child's shape sorter using card.
Use the work you have done in this chapter to help you.

Your project needs careful planning before you start.
Here are some of the things you should consider.

- How many shapes are there going to be?
- How big should each shape be?
- What shapes are you going to have?
- Can you construct the nets of your shapes?
- How are you going to make sure that each shape will only fit through one hole?
- Are all the shapes going to be symmetrical?
- You do not need to make a whole posting box.
  You can just make the lid with the posting holes.
  If you do decide to make a box, make sure it is big enough to contain all the shapes.

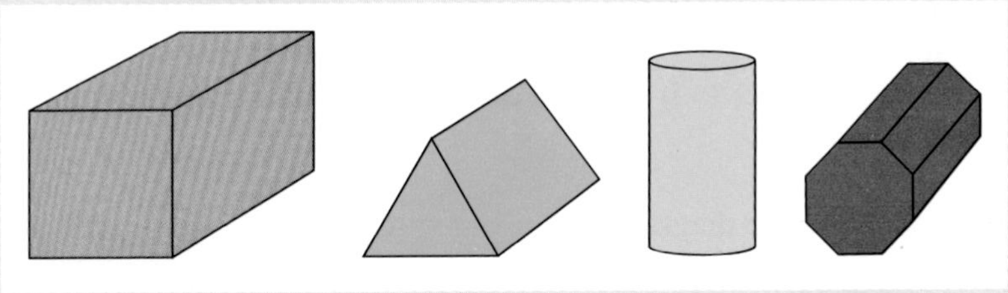

1 The exterior angle of a regular polygon is 18°.
  a How many sides has it got?
  b Write down the size of an interior angle of the polygon.

2 Find the missing angle
in this irregular pentagon.

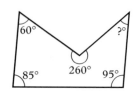

3 Work out the angles marked
in this regular pentagon.

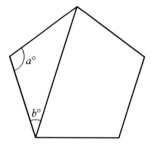

4 A regular polygon has n sides.
  a Write down a formula to work out the exterior angle.
  b Use your answer to **a** to write down a formula for the interior
    angle.
  c Use your answer to **b** to write down a formula for the sum of the
    interior angles.

5 Copy each of these diagrams.
Complete them so that:
  a the dotted line is a
    line of symmetry,
  b C is the centre of rotational
    symmetry for a shape that has
    rotational symmetry of order 3.

6 A regular polygon has 10 sides.
  a How many lines of symmetry
    does the polygon have?
  b What is the order of rotational
    symmetry of the polygon?

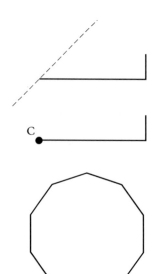

7 Two identical square based pyramids are stuck together to make this octahedron. Write down the number of planes of symmetry of the octahedron.

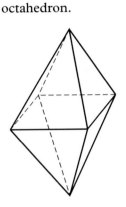

8 Make **two** copies of the diagram on squared paper.
Shade one extra square on each copy so that your new shape has:
a just one line of symmetry but no rotational symmetry,
b no lines of symmetry but rotational symmetry of order 2.

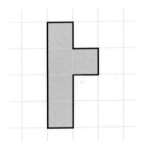

9 Val has drawn this net of a cube. Val folds up the net.
a The tab on edge LK will stick to edge JK.
Write down the letters of the edge that the tab on edge AB will stick to.
b What other two letters will meet point E?
c Make a copy of Val's net.
Make the sides of each square 2 cm long.
Cut out the net and fold it up.
Use your net to check your answers to parts **a** and **b**.

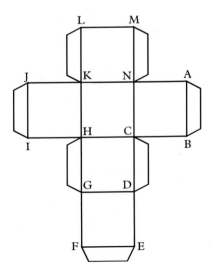

347

1　Here are two semi-regular tessellations.

a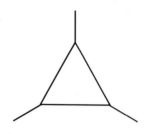

Write down the name of the regular polygon that tessellates with the equilateral triangle.

b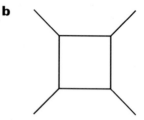

Write down the name of the regular polygon that tessellates with the square.

2　The diagram shows part of a regular polygon. Work out the number of sides of the polygon.

165°

3　Write down the number of planes of symmetry of a cube.

4　Here is an unusual net of a common solid. The net has flaps. The net is folded to make the solid.
　　a　Write down the name of the solid.
　　b　Write down the edges that would stick to these:
　　　　(1) the flap on edge AB　　　　(2) the flap on edge CD
　　c　Which other letter would meet J?

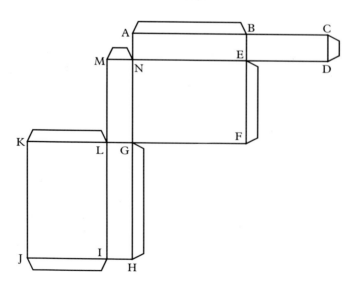

- **Exterior and interior angles**

  The sum of the exterior angles of any polygon is 360°.

  interior angle = 180° − exterior angle

  Example Work out the interior angle of a regular hexagon.

  $$\text{Exterior angle} = \frac{360°}{6} = 60°$$

  Interior angle = 180° − 60° = 120°

- **Sum of the interior angles**

  You can find the **sum of the interior angles** of any polygon.

  Example Find the sum of the interior angles of an octagon.

  The octagon splits into 6 triangles.
  Total of the interior angles = 6 × 180°
  = 1080°

  If it is a regular octagon, each interior angle $= \dfrac{1080°}{8} = 135°$

- **Plane of symmetry**

  A **plane of symmetry** divides a *solid* into two equal parts. Each part is a reflection of the other.

  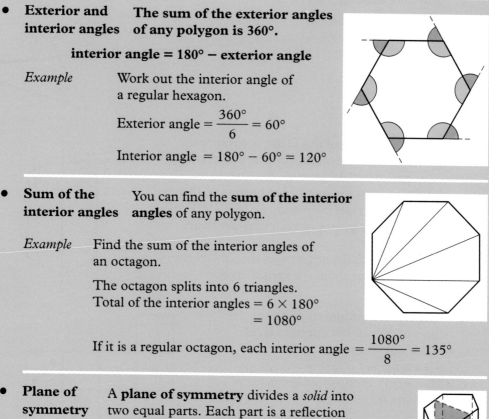

- **Net** A **net** is a pattern of shapes on a piece of paper or card. The shapes are arranged so that the net can be folded to make a hollow solid.
  Half the edges of these nets have flaps. The flaps are on alternate edges.

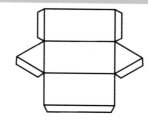
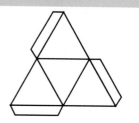
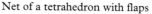

Net of a triangular prism with flaps   Net of a tetrahedron with flaps

1 Calculate the angles marked
   with letters in this polygon.

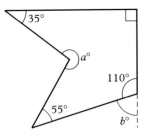

2 The cross section of this prism is a rhombus.
   a Write down the number of vertical planes of
        symmetry of the prism.
   b Write down the number of horizontal planes
        of symmetry of the prism.
   c How many planes of symmetry are there
        altogether?
   d Write down the order of rotational symmetry
        of the prism about the vertical axis.

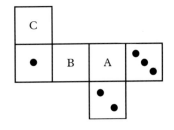

3 The rule for putting spots on the
   faces of dice is that opposite faces
   add up to seven.
   This is the net of a dice.
   How many spots would go on faces
   A, B and C?

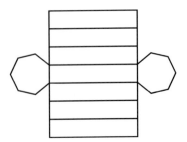

4 Write down the names of the solids that can be made from these nets.
   a   **b**

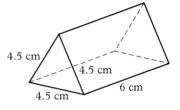

5 Construct an accurate net of this solid.

4.5 cm
4.5 cm
4.5 cm
6 cm

Help yourself

1 Multiplying by 10

When you multiply by 10, all the digits move across **one** column to the **left**. This makes the number 10 times bigger.

You can use the headings **Th H T U** to help.

They mean **Th**ousands, **H**undreds, **T**ens and **U**nits. Units is another way of saying 'ones'.

Example

23 × 10 = 230

H	T	U
	2	3
2	3	0

Here are some more examples:

Th	H	T	U
	4	6	
	4	6	0

46 × 10 = 460

	2	5	3
2	5	3	0

253 × 10 = 2530

Exercise 1

Multiply each of these numbers by 10

1 48   **3** 842   **5** 7000

2 54   **4** 777   **6** 9003

2 Multiplying by 100, 1000, ...

When you multiply by 100, all the digits move across **two** columns to the **left**.

This makes the number 100 times bigger.
This is because 100 = 10 × 10
So multiplying by 100 is like multiplying by 10 twice.

Example

74 × 100 = 7400

Th	H	T	U
		7	4
7	4	0	0

When you multiply by 1000 all the numbers move across three columns to the left.

This is because 1000 = 10 × 10 × 10
This means that multiplying by 1000 is like multiplying by 10 three times.

Example

74 × 1000 = 74 000

TTh	Th	H	T	U
			7	4
7	4	0	0	0

Exercise 2

Write down the answers to these.

1 27 × 100   **7** 4153 × 100

2 91 × 100   **8** 900 × 1000

3 74 × 1000   **9** 4004 × 1000

4 291 × 100   **10** 924 × 10 000

5 4270 × 100   **11** 301 × 10 000

6 840 × 1000   **12** 737 × 100 000

3 Multiplying by 20, 30, ...

When you multiply by 20 it is like multiplying by 2 then by 10. This is because $20 = 2 \times 10$

Example

To do 18×20:
first do

$$
\begin{array}{r}
1\,8 \\
\times \quad 2 \\
\hline
3\,6 \\
{\scriptstyle 1}
\end{array}
$$

Then do $\qquad 36 \times 10 = 360$

So $\qquad 18 \times 20 = 360$

In the same way multiplying by 30 is the same as multiplying by 3 and then multiplying by 10

Example

To do 26×30:
first do

$$
\begin{array}{r}
2\,6 \\
\times \quad 3 \\
\hline
7\,8 \\
{\scriptstyle 1}
\end{array}
$$

Then do $\qquad 78 \times 10 = 780$

So $\qquad 26 \times 30 = 780$

Exercise 3

Work these out.

1	39×20	**7**	92×40
2	42×20	**8**	25×50
3	26×30	**9**	71×50
4	23×30	**10**	304×20
5	65×30	**11**	291×30
6	34×40	**12**	525×70

4 Multiplying decimals by 10

You can multiply decimals by 10 in the same way.

Examples

1 41.5×10

H T U . $\frac{1}{10}$

$4 \quad 1 \,. \, 5$ $\qquad 41.5 \times 10 = 415$
$4 \quad 1 \quad 5 \,. \, 0$

2 56.87×10

H T U . $\frac{1}{10}$ $\frac{1}{100}$

$5 \quad 6 \,. \, 8 \quad 7$ $\qquad 56.87 \times 10$
$5 \quad 6 \quad 8 \,. \, 7$ $\qquad\qquad = 568.7$

Exercise 4

Multiply these decimals by 10

1	7.4	**4**	72.34
2	32.5	**5**	20.8
3	18.91	**6**	0.4

353

5 Multiplying decimals by 100

When you multiply by 100, all the digits move across **two** columns to the **left**.

Examples

1 27.65×100

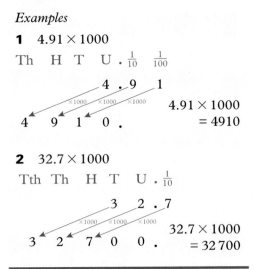

Th	H	T	U .	$\frac{1}{10}$	$\frac{1}{100}$
		2	7 .	6	5
2	7	6	5 .		

27.65×100
$= 2765$

2 96.5×100

Th	H	T	U .	$\frac{1}{10}$
	9	6 .	5	
9	6	5	0 .	

96.5×100
$= 9650$

Exercise 5

Multiply these decimals by 100

1 65.86 **4** 721.8

2 22.94 **5** 70.39

3 16.4 **6** 4.01

6 Multiplying decimals by 1000

When you multiply by 1000, all the digits move across **three** columns to the **left**.

Examples

1 4.91×1000

Th	H	T	U .	$\frac{1}{10}$	$\frac{1}{100}$
			4 .	9	1
4	9	1	0 .		

4.91×1000
$= 4910$

2 32.7×1000

Tth	Th	H	T	U .	$\frac{1}{10}$
			3	2 .	7
3	2	7	0	0 .	

32.7×1000
$= 32\,700$

Exercise 6

Multiply these decimals by 1000

1 21.46 **4** 56.89

2 78.91 **5** 99.04

3 5.24 **6** 6.01

7 Long multiplication

When you want to multiply two quite large numbers you have to do it in stages. Here are two methods. You only have to know one of them.

Method 1

Example
146 × 24

First do 146 × 4

```
    1 4 6
×       4
    5 8 4
    1 2
```

Then do 146 × 20

```
    1 4 6
×       2
    2 9 2
      1
```

292 × 10 = 2920

Now add the two answers together.

```
    5 8 4
+ 2 9 2 0
  3 5 0 4
```

Usually the working out looks like this:

```
    1 4 6
×     2 4
    5 8 4
  2 9 2 0
  3 5 0 4
  1 1
```

Here is another example.

```
      2 2 3
×       3 6
  1 3 3 8  ← (223 × 6)
  6 6 9 0  ← (223 × 30)
  8 0 2 8
  1 1
```

Method 2

Example
125 × 23

First set out the numbers with boxes, like this:

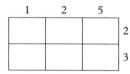

Now draw in the diagonals like this:

Fill in like a table square then add along the diagonals like this:

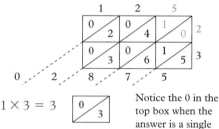

1 × 3 = 3   Notice the 0 in the top box when the answer is a single digit.

So 125 × 23 = **2875**

Here is another example.
When the diagonal adds up to more than 10, you carry into the next one.

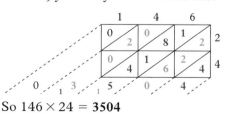

So 146 × 24 = **3504**

Exercise 7

Use the method you prefer to work these out.

1 27×25

2 76×24

3 123×53

4 404×26

5 382×25

6 271×54

7 391×45

8 317×84

9 545×22

10 821×65

11 754×71

12 989×89

8 Long Division

Sometimes you need to do long division. This is usually when you are dividing by a number bigger than 10

Example

$468 \div 12$

$$12\overline{)468}$$

12 will not go into 4 so

first do $46 \div 12$

You need to find out how many times 12 goes into 46

$$12 \times 2 = 24$$
$$12 \times 3 = 36 \leftarrow$$
$$12 \times 4 = 48$$

12 will go in 3 times.
Put the 3 above the 6

$$12\overline{)468} \quad \overset{3}{}$$

Work out 3×12 and put the answer under the 46

$$\begin{array}{r} 3 \\ 12\overline{)468} \\ 36 \end{array}$$

Now subtract the 36 from the 46

$$\begin{array}{r} 3 \\ 12\overline{)468} \\ \underline{36} \\ 10 \end{array}$$

The 10 is the carry.
You don't put it with the 8
Instead you bring the 8 down to the 10

$$\begin{array}{r} 3 \\ 12\overline{)468} \\ 36\downarrow \\ \hline 108 \end{array}$$

Now do $108 \div 12$

$$12 \times 8 = 96$$
$$12 \times 9 = 108 \leftarrow$$

12 will go in 9 times exactly.
Put the 9 after the 3

$$\begin{array}{r} 39 \\ 12\overline{)468} \\ 36 \\ \hline 108 \end{array}$$

Work out 9×12 and put the answer under the 108

When you subtract this time there is no remainder.

You have finished!

$$\begin{array}{r} 39 \\ 12\overline{)468} \\ 36 \\ \hline 108 \\ \underline{108} \\ - \end{array}$$

So $468 \div 12 = 39$

Exercise 8

Work these out.

1	$540 \div 12$	**7**	$805 \div 23$
2	$806 \div 13$	**8**	$754 \div 26$
3	$938 \div 14$	**9**	$928 \div 32$
4	$690 \div 15$	**10**	$648 \div 27$
5	$600 \div 12$	**11**	$1056 \div 16$
6	$540 \div 18$	**12**	$1656 \div 18$

Sometimes there is a remainder left at the end.

Example

$383 \div 14$

$$
\begin{array}{r}
2\,7 \\
14\overline{)383} \\
2\,8 \\
\hline
10\,3 \\
9\,8 \\
\hline
5
\end{array}
$$

You could carry on and divide the 5 by the 14 to get a decimal.

It is easier to leave it as a fraction.

$5 \div 14$ is the fraction $\frac{5}{14}$

So $383 \div 14 = 27\frac{5}{14}$

Exercise 9

Work these out.

1	$698 \div 12$	**7**	$212 \div 14$
2	$664 \div 13$	**8**	$517 \div 17$
3	$818 \div 16$	**9**	$209 \div 13$
4	$925 \div 22$	**10**	$310 \div 18$
5	$550 \div 24$	**11**	$2834 \div 14$
6	$872 \div 32$	**12**	$8721 \div 15$

9 Dividing by 10

When you divide by 10, all the digits move across **one** column to the **right**. This makes the number smaller.

Example

$230 \div 10 = 23$

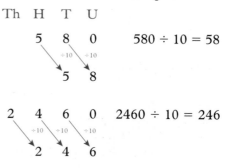

Here are some more examples.

Th H T U

$580 \div 10 = 58$

$2460 \div 10 = 246$

Exercise 10

Divide these numbers by 10

1	820	**5**	8160
2	60	**6**	9400
3	4820	**7**	7000
4	930	**8**	500 000

10 Dividing by 100, 1000, ...

When you divide by 100, all the digits move across **two** columns to the **right**. This is because $100 = 10 \times 10$
So dividing by 100 is like dividing by 10 twice.

357

Example

$7400 \div 100 = 74$

Th H T U

7   4   0   0

$\div 100$ $\div 100$

7   4

When you divide by 1000, all the numbers move across **three** columns to the **right**.

Example

$74\,000 \div 1000 = 74$

TTh Th H T U

7   4   0   0   0

$\div 1000$ $\div 1000$

7   4

Exercise 11

Work these out.

1 $5400 \div 100$

2 $7100 \div 100$

3 $8200 \div 100$

4 $64\,000 \div 1000$

5 $84\,000 \div 100$

6 $84\,000 \div 1000$

7 $400\,000 \div 1000$

8 $400\,000 \div 10\,000$

11 Dividing by 20, 30, ...

When you divide by 20, it is like dividing by 2 then by 10. This is because $20 = 2 \times 10$

Example

To do $360 \div 20$

first do

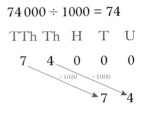

Then do $180 \div 10 = 18$

So $360 \div 20 = 18$

In the same way dividing by 30 is the same as dividing by 3 then by 10

Example

To do $780 \div 30$

first do
$$\frac{260}{3\overline{)7\,{}^18\,0}}$$

Then do $260 \div 10 = 26$

So $780 \div 30 = 26$

Exercise 12

Work these out.

1 $820 \div 20$

2 $480 \div 30$

3 $3720 \div 40$

4 $5250 \div 50$

5 $7520 \div 20$

6 $4620 \div 30$

7 $1980 \div 90$

8 $24\,480 \div 80$

12 Dividing decimals by 10

You can divide decimals by 10 in the same way.

Examples

1 $47.1 \div 10$

T U . $\frac{1}{10}$ $\frac{1}{100}$

4 7 . 1

$\div 10$ $\div 10$ $\div 10$

4 . 7 1

$47.1 \div 10 = 4.71$

2 $2.9 \div 10$

U . $\frac{1}{10}$ $\frac{1}{100}$

2 . 9

$\div 10$ $\div 10$

0 . 2 9

$2.9 \div 10 = 0.29$

Exercise 13

Divide these decimals by 10

1	32.7	**3**	3.4	**5**	3.01
2	96.4	**4**	8.79	**6**	10.3

13 Dividing decimals by 100

When you divide by 100, all the digits move across **two** columns to the **right**.

Examples

1 257.1 ÷ 100

H T U . $\frac{1}{10}$ $\frac{1}{100}$ $\frac{1}{1000}$

2   5   7 . 1   257.1 ÷ 100
    ÷100 ÷100 ÷100 ÷100   = 2.571
      2 . 5   7   1

2 52.3 ÷ 100

T U . $\frac{1}{10}$ $\frac{1}{100}$ $\frac{1}{1000}$

5   2 . 3   52.3 ÷ 100
    ÷100 ÷100 ÷100   = 0.523
  0 . 5   2   3

Exercise 14

Divide these decimals by 100

1	182.5	**3**	23.4	**5**	10.2
2	479.1	**4**	17.6	**6**	31.02

Other words

These words can also mean **divide**.

share   **quotient**

Examples

Share 240 by 12 ⎫
Find the **quotient** ⎬ both mean
of 240 and 12 ⎭ 240 ÷ 12

14 Dividing decimals by 1000

When you divide by 1000, all the digits move across **three** columns to the **right**.

Examples

1 342.6 ÷ 1000

H   T   U . $\frac{1}{10}$ $\frac{1}{100}$ $\frac{1}{1000}$ $\frac{1}{10\,000}$

3   4   2 . 6
    ÷1000 ÷1000 ÷1000 ÷1000
        0 . 3   4   2   6

342.6 ÷ 1000 = 0.3426

2 75.1 ÷ 1000

H   T   U . $\frac{1}{10}$ $\frac{1}{100}$ $\frac{1}{1000}$ $\frac{1}{10\,000}$

7   5 . 1
    ÷1000 ÷1000 ÷1000
    0 . 0   7   5   1

Exercise 15

Divide these decimals by 1000

1	691.4	**3**	16.7	**5**	6.71
2	872.6	**4**	34.2	**6**	8.04

15 Adding fractions

To add fractions, the bottom numbers (denominators) **must** be the same.

Examples

$\frac{2}{7} + \frac{3}{7} = \frac{5}{7}$

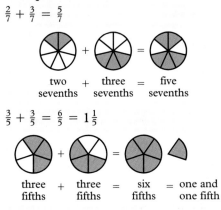

two + three = five
sevenths sevenths sevenths

$\frac{3}{5} + \frac{3}{5} = \frac{6}{5} = 1\frac{1}{5}$

three + three = six = one and
fifths fifths fifths one fifth

Exercise 16

Work these out.

1 $\frac{3}{9} + \frac{4}{9}$ **3** $\frac{9}{12} + \frac{2}{12}$ **5** $\frac{6}{8} + \frac{7}{8}$

2 $\frac{6}{12} + \frac{5}{12}$ **4** $\frac{4}{5} + \frac{4}{5}$ **6** $\frac{9}{11} + \frac{8}{11}$

Sometimes the two bottom numbers are different. Before you can add the fractions you **must** make them the same.

Example

$\frac{2}{3} + \frac{1}{6}$

You need to find a number that 3 and 6 both divide into exactly.

Numbers that 3 goes into:
3 ⑥ 9 12 ...
Numbers that 6 goes into:
⑥ 12 18 ...

The first number that is in both lists is 6. The 6 is called the common denominator.

360

Now write the fractions with 6 as the bottom number:

$\frac{2}{3} = \frac{?}{6}$ so $\frac{2}{3}$ $\frac{4}{6}$ so $\frac{2}{3} = \frac{4}{6}$.

You can see this in a diagram.

The $\frac{1}{6}$ does not need changing.

So $\frac{2}{3} + \frac{1}{6} = \frac{4}{6} + \frac{1}{6} = \frac{5}{6}$

Here is another example

$\frac{2}{3} + \frac{1}{4}$

Numbers that 3 goes into:
3 6 9 ⑫ 15 ...

Numbers that 4 goes into:
4 8 ⑫ 16 ...

You need to change both fractions to twelfths. 12 is the common denominator.

$\frac{2}{3} = \frac{?}{12}$ $\frac{2}{3} = \frac{8}{12}$

$\frac{1}{4} = \frac{?}{12}$ $\frac{1}{4} = \frac{3}{12}$

So $\frac{2}{3} + \frac{1}{4} = \frac{8}{12} + \frac{3}{12} = \frac{11}{12}$

Exercise 17

Work these out.

1 $\frac{1}{4} + \frac{1}{8}$ **7** $\frac{1}{3} + \frac{1}{4}$

2 $\frac{1}{5} + \frac{1}{15}$ **8** $\frac{2}{7} + \frac{1}{3}$

3 $\frac{2}{7} + \frac{3}{14}$ **9** $\frac{2}{5} + \frac{1}{6}$

4 $\frac{5}{9} + \frac{1}{3}$ **10** $\frac{1}{7} + \frac{4}{8}$

5 $\frac{5}{9} + \frac{1}{18}$ **11** $\frac{1}{8} + \frac{3}{5}$

6 $\frac{1}{3} + \frac{1}{6}$ **12** $\frac{1}{2} + \frac{1}{3} + \frac{1}{4}$

16 Subtracting fractions

This works just like adding fractions.

Example

$\frac{3}{5} - \frac{2}{5} = \frac{1}{5}$

The two bottom numbers must still be the same.

Example

$\frac{3}{8} - \frac{1}{4}$

Numbers that 8 goes into:
⑧ 16 24 …
Numbers that 4 goes into:
4 ⑧ 12 16 …

$\frac{1}{4} = \frac{?}{8}$ $\frac{1}{4} \overset{\times 2}{=} \frac{2}{8}$
  ↘×2   ↘×2

The $\frac{3}{8}$ does not need changing.

So $\frac{3}{8} - \frac{1}{4} = \frac{3}{8} - \frac{2}{8} = \frac{1}{8}$

17 Simplifying fractions

This is also known as **cancelling**.

You look for a number that divides exactly into both the top and bottom numbers.

Examples

1 Simplify $\frac{6}{15}$

3 goes into both 6 and 15 exactly

$\frac{6}{15} \overset{\div 3}{\underset{\div 3}{=}} \frac{2}{5}$

You can divide by more than one number.

2 Simplify $\frac{18}{24}$

$\frac{18}{24} \overset{\div 2}{\underset{\div 2}{=}} \frac{9}{12} \overset{\div 3}{\underset{\div 3}{=}} \frac{3}{4}$

Exercise 18

Work these out.

1 $\frac{5}{8} - \frac{2}{8}$ **7** $\frac{1}{4} - \frac{1}{6}$

2 $\frac{3}{5} - \frac{2}{5}$ **8** $\frac{2}{4} - \frac{1}{3}$

3 $\frac{6}{11} - \frac{2}{11}$ **9** $\frac{7}{8} - \frac{2}{3}$

4 $\frac{2}{5} - \frac{1}{10}$ **10** $\frac{3}{4} - \frac{1}{3}$

5 $\frac{6}{8} - \frac{1}{4}$ **11** $\frac{5}{8} - \frac{2}{6}$

6 $\frac{11}{12} - \frac{1}{3}$ **12** $\frac{10}{11} - \frac{6}{8}$

Exercise 19

Simplify these.

1 $\frac{2}{6}$ **7** $\frac{24}{36}$

2 $\frac{4}{12}$ **8** $\frac{18}{27}$

3 $\frac{5}{15}$ **9** $\frac{25}{35}$

4 $\frac{6}{9}$ **10** $\frac{8}{40}$

5 $\frac{8}{12}$ **11** $\frac{30}{60}$

6 $\frac{30}{50}$ **12** $\frac{38}{80}$

18 Converting units

Common metric units of length

10 millimetres (mm) = 1 centimetre (cm)
100 centimetres = 1 metre (m)
1000 metres = 1 kilometre (km)

Examples

1 Convert 6.9 cm to mm.
6.9 cm = 6.9 × 10 mm
= 69 mm

2 Convert 5.34 m to cm.
5.34 m = 5.34 × 100 cm
= 5.34 cm

3 Convert 7.3 km to m.
7.3 km = 7.3 × 1000 m
= 7300 m

Exercise 20

1 Convert these lengths to mm.
a 3.4 cm   **c** 131 cm
b 12.8 cm   **d** 113.7 cm

2 Convert these lengths to cm.
a 4.1 m   **c** 12.1 m
b 2.8 m   **d** 324 m

3 Convert these lengths to m.
a 8.8 km   **c** 15 km
b 9.7 km   **d** 100 km

Common metric units of mass

Units of mass have similar names to units of length.

1000 milligrams (mg) = 1 gram (g)
1000 grams = 1 kilogram (kg)
1000 kg = 1 tonne (t)

Examples

1 Convert 2.5 kg to g.
2.5 kg = 2.5 × 1000 g
= 2500 g

2 Convert 5000 g to kg.
5000 g = 5000 ÷ 1000 kg
= 5 kg

Exercise 21

Convert the units in each of these.
Think carefully whether you need to
multiply or divide.

1 **a** 3 kg to g   **d** 3000 g to kg
   b 7 kg to g   **e** 6000 g to kg
   c 21 kg to g   **f** 800 g to kg

2 **a** 3.5 kg to g   **d** 6500 g to kg
   b 7.2 kg to g   **e** 3200 g to kg
   c 3.84 kg to g   **f** 2800 g to kg

3 **a** 5 g to mg   **d** 3000 kg to t
   b 8000 mg to g   **e** 4 t to kg
   c 1640 mg to g   **f** 2.5 t to kg

CHAPTER 1

1 **a** $s^2 = 4^2 + 5^2$
$s^2 = 41$
$s = \sqrt{41}$
$= 6.4$ cm correct to 1 dp

    b $18^2 = 14^2 + n^2$
$n^2 + 196 = 324$
$n^2 = 128$
$n = 11.3$ cm correct to 1 dp

    c $56^2 = y^2 + 45^2$
$y^2 + 2025 = 3136$
$y^2 = 1111$
$y = 33.3$ mm correct to 1 dp

    d 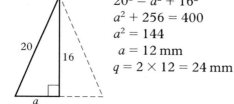 $20^2 = a^2 + 16^2$
$a^2 + 256 = 400$
$a^2 = 144$
$a = 12$ mm
$q = 2 \times 12 = 24$ mm

2 $8.5^2 = 72.25$ (This is the longest side.)
$4^2 + 7.5^2 = 16 + 56.25 = 72.25$
The triangle has a right angle.

3 $3.5^2 = h^2 + 1.2^2$
$h^2 + 1.2^2 = 3.5^2$
$h^2 = 12.25 - 1.44$
$h = 3.29$ correct to 2 dp

CHAPTER 2

1 **a** $T = 2 \times n + 70 = 2n + 70$
    b $T = 2 \times 54 + 70 = £178$

2 **a** $D = \dfrac{14.8^2 + 12.3^2}{14.8 - 12.3} = 148.1$ to 1 dp
    b $D = 3\sqrt{19^2 + 14^2} = 70.8$ to 1 dp

3 **a**

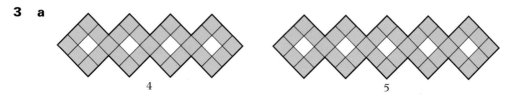

4   5

b

Number of pattern	1	2	3	4	5
Number of tiles	8	15	22	29	36

c 7

d $25 \times 7 + 1 = 176$

e $t = 7n + 1$

4 a $t = 7n + 1$

$t = 7 \times 6 + 1 = 43$ correct

b

The formula is $2n^2 + \ldots n + \ldots$

Number of term	1	2	3	4	5
Pattern	9	19	33	51	73
$2n^2$	2	8	18	32	50
Rest of pattern	7	11	15	19	23

The formula for the 'rest of pattern' is $4n + 3$

The final formula is $2n^2 + 4n + 3$

Check to find term when $n = 5$:

$2n^2 + 4n + 3 = 50 + 20 + 3 = 73$ correct

5

Value of x	Value of $x^2 + 2x$	
5	35	too small
6	48	too big
5.8	45.24	too small
5.9	46.61	too big
5.85	45.9225	too small
5.86	46.0596	too big
5.855	45.9910	too small

x lies between 5.855 and 5.86

$x = 5.86$ to 2 dp.

6 Value of x Value of $x(14 + x) = 45$

Value of x	Value of $x(14+x)=45$	
2	32	too small
3	51	too big
2.5	41.25	too small
2.6	43.16	too small
2.7	45.09	too big

x lies between 2.6 and 2.7

CHAPTER 3

1 **a** $C = \pi \times d$
$= \pi \times 18$
$C = 56.5$ cm to 1 dp

    b $d = 2 \times r$
$= 2 \times 4.1$
$d = 8.2$ cm

$C = \pi \times d$
$= \pi \times 8.2$
$C = 25.8$ cm to 1 dp

2 $d = \dfrac{C}{\pi}$

$= \dfrac{82}{\pi}$

$= 26.101\ 411$ [Keep this number on your calculator to use in the next part]

$r = d \div 2$
$= 26.101\ 411 \div 2$
$r = 13.1$ cm to 1 dp   [You only round at the end of a question]

3 **a** $A = \pi r^2$
$= \pi \times 2.9^2$
$A = 26.4$ cm^2 to 1 dp

    b $r = d \div 2$
$= 30 \div 2$
$r = 15$ cm

$A = \pi r^2$
$= \pi \times 15^2$
$A = 706.9$ cm^2 to 1 dp

4 a This shape is one quarter of a circle.

Area of whole circle
$$= \pi r^2$$
$$= \pi \times 7^2$$

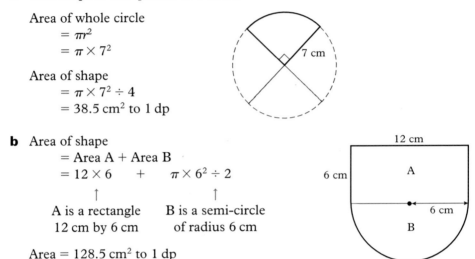

7 cm

Area of shape
$$= \pi \times 7^2 \div 4$$
$$= 38.5 \text{ cm}^2 \text{ to 1 dp}$$

b Area of shape
$$= \text{Area A} + \text{Area B}$$
$$= 12 \times 6 \quad + \quad \pi \times 6^2 \div 2$$

↑   ↑

A is a rectangle   B is a semi-circle
12 cm by 6 cm   of radius 6 cm

Area $= 128.5 \text{ cm}^2$ to 1 dp

12 cm

A

6 cm

6 cm

B

5 The path is coloured red on this diagram.
The flower bed is green.

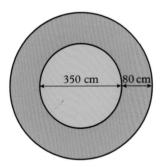

350 cm   80 cm

Area of path = Area of large circle − Area of flower bed

Radius of flower bed $= 350 \div 2$
$$= 175 \text{ cm}$$

Radius of large circle $= 175 + 80$
$$= 255 \text{ cm}$$

Area of path $= \pi \times 255^2 - \pi \times 175^2$
$$= 108\,070.79 \text{ cm}^2$$
$$= 10.8 \text{ m}^2 \text{ to 1 dp}$$

$(1 \text{ m}^2 = 100 \times 100 = 10\,000 \text{ cm}^2$ so to go from cm^2 to $\text{m}^2 \div$ by $10\,000)$

CHAPTER 4

1 a

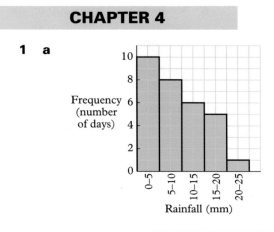

b April is wetter. The bar-chart for April has the taller bars on the right, ie there are more days with higher rainfall than November, where the tall bars are on the left for small amounts of rain.

c November:

Rainfall (midpoint)	Number of days	Amount of rain	
2.5	10	25	(10×2.5)
7.5	8	60	(8×7.5)
12.5	6	75	(6×12.5)
17.5	5	87.5	(5×17.5)
22.5	1	22.5	(1×22.5)
	Total 30	Total 270	

$$\text{Estimate of mean rainfall} = \frac{270}{30} = 9\,\text{mm}$$

2 a

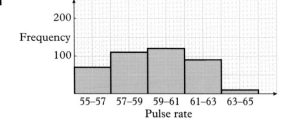

b

Pulse rate (mid point)	Frequency	Mid point × frequency
56	70	3920
58	110	6380
60	120	7200
62	90	5580
64	10	640
Total 400	Total 23 720	

c

Pulse rate below	Cumulative frequency
57	70
59	180
61	300
63	390
65	400

$$\text{Estimate of mean pulse rate} = \frac{23\,720}{400} = 59.3$$

d

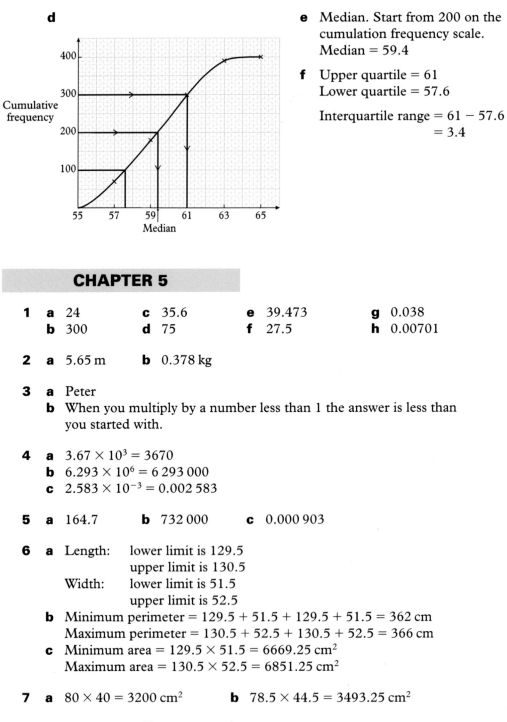

e Median. Start from 200 on the cumulation frequency scale.
Median = 59.4

f Upper quartile = 61
Lower quartile = 57.6

Interquartile range = 61 − 57.6
= 3.4

CHAPTER 5

1 **a** 24 **c** 35.6 **e** 39.473 **g** 0.038
  b 300 **d** 75 **f** 27.5 **h** 0.00701

2 **a** 5.65 m **b** 0.378 kg

3 **a** Peter
  b When you multiply by a number less than 1 the answer is less than you started with.

4 **a** $3.67 \times 10^3 = 3670$
  b $6.293 \times 10^6 = 6\,293\,000$
  c $2.583 \times 10^{-3} = 0.002\,583$

5 **a** 164.7 **b** 732 000 **c** 0.000 903

6 **a** Length: lower limit is 129.5
    upper limit is 130.5
   Width: lower limit is 51.5
    upper limit is 52.5
  b Minimum perimeter = 129.5 + 51.5 + 129.5 + 51.5 = 362 cm
   Maximum perimeter = 130.5 + 52.5 + 130.5 + 52.5 = 366 cm
  c Minimum area = $129.5 \times 51.5 = 6669.25$ cm^2
   Maximum area = $130.5 \times 52.5 = 6851.25$ cm^2

7 **a** $80 \times 40 = 3200$ cm^2 **b** $78.5 \times 44.5 = 3493.25$ cm^2

8 Time = $1.5 \times 10^{11} \div 1.99 \times 10^4$ m
   = $7.537\,688 \times 10^6$

CHAPTER 6

1 Volume = 3.9×4.1 cm^3 = 15.99 cm^3

2 **a** Area of cross section = $\dfrac{15 + 9}{2} \times 8 = 12 \times 8 = 96$ cm^2
Volume = $96 \times 20 = 1920$ cm^3
b Area of cross section = $3.14 \times 12 \times 12 = 452.16$ cm^2
Volume = 452.16×20 cm^3 = 9043.2 cm^3.

3 Area of cross section = $3.14 \times 100 \times 100 = 31\,400$
Volume = $31\,400 \times 30 = 942\,000$
1 litre = 1000 ml = 1000 cm^3
$\therefore$ Volume = $942\,000 \div 1000 \, l = 942 \, l$

4 **a** Volume = $2.1 \times 0.6 \times 0.8 = 1.008$ cm^3
b Volume = $21 \times 6 \times 8 = 1008$ mm^3

5 **a** Perimeter: $4p - kr + 2s$
b Area: $\pi q^2 + 4pr$; $kpq + 3\pi pq$; $kp^2 + \pi s^2$; $7kpq - 5s^2$
c Volume: $pqr - kp^3$

CHAPTER 7

1 **a** $\frac{4}{5} \times 100 = 4 \div 5 \times 100 = 80\%$
b $0.39 \times 100 = 39\%$

2 **a** $16\% = 16 \div 100 = 0.16$
b $\frac{16}{100} = \frac{4}{25}$

3 5% of $240 = \frac{5}{100} \times 240 = £12$
His new salary is
$£240 + £12 = £252$

4 $100\% - 12\frac{1}{2}\% = 87.5\%$
87.5% is £367.50
1% is $£367.50 \div 87.5 = £4.20$
100% is $£4.20 \times 100 = £420$
Original price is £420

5 $5 + 3 + 1 = 9$ shares are needed
One share is $£360 \div 9 = £40$
1st prize is $£40 \times 5 = £200$
2nd prize is $£40 \times 3 = £120$
3rd prize is $£40 \times 1 = £40$
Check: $£200 + £120 + £40 = £360$

6 **a** $5 : 12 = \frac{5}{5} : \frac{12}{5} = 1 : 2.4$
b $5 : 12 = \frac{5}{12} : \frac{12}{12} = 0.42 : 1$

7 1 car costs $£2.61 \div 3 = £0.87$
7 cars cost $£0.87 \times 7 = £6.09$

8　**a**　Density of mercury is $67.7 \div 5 = 13.54 \text{ g/cm}^3$
　　b　Mass of 12 cm^3 of mercury is $13.54 \times 12 = 162.48 \text{ g}$

9　Time, $\text{T} = \dfrac{\text{D}}{\text{S}} = 2320 \div 580 = 4 \text{ hours}$

10　Distance $\text{D} = \text{S} \times \text{T} = 300\,000 \times 6 \text{ km} = 1\,800\,000 \text{ km}$

11　Volume $\text{V} = \dfrac{\text{M}}{\text{D}} = 563.5 \div 2.3 = 245 \text{ g}$

CHAPTER 8

1　**a**　0　　　　　**b**　$5k - 13$　　　　　**c**　$-r - s$

2　**a**　y^4　　　　　**b**　$-12r^2$　　　　　**c**　$15t^3$

3　**a**　$-12 + 16r$　　**b**　$12 - 9g + 35 - 10g = 47 - 19g$

4　**a**　$(2g - h)(g + 3h) = 2g(g + 3h) - h(g + 3h)$
　　　　　　　　　　　　　　$= 2g^2 + 6gh - hg - 3h^2$
　　　　　　　　　　　　　　$= 2g^2 + 5gh - 3h^2$
　　b　$(2r - 1)(2r + 1) = (2r)^2 - 1^2$
　　　　　　　　　　　　$= 4r^2 - 1$
　　c　$(3 + f)(7 - 2f) = 3(7 - 2f) + f(7 - 2f)$
　　　　　　　　　　　　$= 21 - 6f + 7f - 2f^2$
　　　　　　　　　　　　$= 21 + f - 2f^2$
　　d　$(5r + 3s)^2 = (5r)^2 + 2 \times 5r \times 3s + (3s)^2$
　　　　　　　　　　$= 25r^2 + 30rs + 9s^2$

5　**a**　$4(d - 3)$　　**b**　$15\,(1 - 2r)$　　**c**　$y(2y + 1)$　　　　**d**　$pq\,(3 - p)$

6　**a**　　　　　$3x + 4 = 19$　　　　　**c**　　　　　$5 + 6x = 3x + 35$
　　　　　　$3x + 4 - 4 = 19 - 4$　　　　　　　$5 + 6x - 3x = 3x - 3x + 35$
　　　　　　　　　　$3x = 15$　　　　　　　　　　　$5 + 3x = 35$
　　　　　　　　　　　$x = 5$　　　　　　　　　$5 - 5 + 3x = 35 - 5$
　　　　　　　　　　　　　　　　　　　　　　　　　　$3x = 30$
　　　　　　　　　　　　　　　　　　　　　　　　　　　$x = 10$

　　b　　　　　$\dfrac{x}{5} - 9 = 2$　　　　　**d**　　　　　$3x - 13 = 35$
　　　　　　　　　　　　　　　　　　　　　　　　$3x - 13 + 13 = 35 + 13$
　　　　　$\dfrac{x}{5} - 9 + 9 = 2 + 9$　　　　　　　　　　$3x = 48$
　　　　　　　　　　　　　　　　　　　　　　　　　　　$x = 16$
　　　　　　　　$\dfrac{x}{5} = 11$

　　　　　　　　　$x = 55$

7 a
$$3x - 11 = 27 - x$$
$$3x + \boldsymbol{x} - 11 = 27 - x + \boldsymbol{x}$$
$$4x - 11 = 27$$
$$4x - 11 + \mathbf{11} = 27 + \mathbf{11}$$
$$4x = 38$$
$$x = 9.5$$

b
$$2x + 14 = 5x - 10$$
$$2x - \boldsymbol{2x} + 14 = 5x - \boldsymbol{2x} - 10$$
$$14 = 3x - 10$$
$$14 + \mathbf{10} = 3x - 10 + \mathbf{10}$$
$$24 = 3x$$
$$8 = x$$
$$\text{or} \quad x = 8$$

8 a
$$6(x - 3) = 24$$
$$6x - 18 = 24$$
$$6x - 18 + \mathbf{18} = 24 + \mathbf{18}$$
$$6x = 42$$
$$x = 7$$

b
$$2(3x + 4) = 38$$
$$6x + 8 = 38$$
$$6x + 8 - \mathbf{8} = 38 - \mathbf{8}$$
$$6x = 30$$
$$x = 5$$

9 a
$$C = mv - mu$$
$$C + \boldsymbol{mu} = mv - mu + \boldsymbol{mu}$$
$$C + mu = mv$$
$$\frac{C + mu}{m} = v$$
$$\text{or} \quad v = \frac{C + mu}{m}$$

b
$$e = \frac{3f - g}{2}$$
$$\mathbf{2 \times} e = \frac{3f - g}{2} \times \mathbf{2}$$
$$2e = 3f - g$$
$$2e + \boldsymbol{g} = 3f - g + \boldsymbol{g}$$
$$2e + g = 3f$$
$$\frac{2e + g}{3} = f$$
$$\text{or} \quad f = \frac{2e + g}{3}$$

10 a
$$E = \tfrac{1}{2}mv^2$$
$$\mathbf{2 \times} E = \mathbf{2 \times} \tfrac{1}{2}mv^2$$
$$2E = mv^2$$
$$\frac{2E}{m} = v^2$$
$$\sqrt{\frac{2E}{m}} = v$$
$$\text{or} \quad v = \sqrt{\frac{2E}{m}}$$

b
$$r = \sqrt{\frac{4s}{9}}$$
square both sides
$$r^2 = \frac{4s}{9}$$
$$\mathbf{9 \times} r^2 = \mathbf{9 \times} \frac{4s}{9}$$
$$9r^2 = 4s$$
$$\frac{9r^2}{4} = s$$
$$\text{or} \quad s = \frac{9r^2}{4}$$

CHAPTER 9

1 a

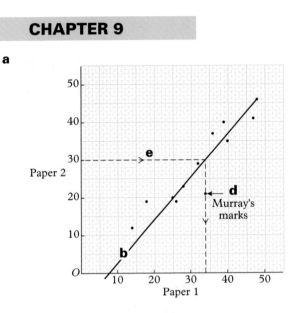

c Paper 1 because the marks are higher.
d 21 (see mark on graph)
e 34 (see dotted line on graph)

2 a $\frac{1}{4}$

b The section for Europe is a bit less than 50%.
An estimate for the percentage is 40%.
(You can give any answer in the range 35% to 45%.)

c $\frac{1}{4}$ of the form chose America
$\frac{1}{4}$ of $28 = \frac{1}{4} \times 28 = 7$ pupils

d Total number of pupils = 30
Each person gets $360° \div 30 = 12°$

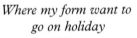

Where my form want to go on holiday

Holiday choice	Number of pupils	Angle
Europe	10	120°
Australia	6	72°
America	8	96°
Africa	2	24°
Asia	4	48°

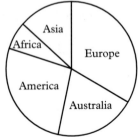

3 The bars are not the same width so the Spring term looks a lot more than 3 times the Autumn term.

1 **a** (1) 0.01   (2) 0.01   (3) 1.00
   b (1) 0.02   (2) 0.01   (3) 1.00

2 **a** $a = 6.9$ cm   **b** $b = 9.8$ cm

3 **a** $a = 56.3°$ **b** $b = 68.7°$

4 $$\sin 75° = \frac{h}{4}$$
$$4 \times \sin 75° = h$$
$$h = 3.86 \text{ m}$$

5 $$\tan ?° = \frac{20}{115}$$
$$? = 10°$$

1 **a** False. There may be more pupils choosing one language than the other.
   b True

2 $1 - 0.8 = 0.2$

3 **a** (1) $\frac{31}{100}$ (2) 0.31   (3) 31%
   b Chicken and chips as this has the highest frequency.

4 **a**

		Dice					
		1	2	3	4	5	6
Spinner	W	W, 1	W, 2	W, 3	W, 4	W, 5	W, 6
	B	B, 1	B, 2	B, 3	B, 4	B, 5	B, 6
	R	R, 1	R, 2	R, 3	R, 4	R, 5	R, 6
	G	G, 1	G, 2	G, 3	G, 4	G, 5	G, 6

   b There are 24 possible outcomes altogether.
   (1) Probability of a red and a 6 $= \frac{1}{24}$
   (2) Probability of a green and a 1 $= \frac{1}{24}$

5

	5 p coin	
	H	T
20 p coin H	H, H	H, T
T	T, H	T, T

There are 4 possible outcomes
1 outcome gives two heads
∴ Probability of getting two heads $= \frac{1}{4}$

6 **a**

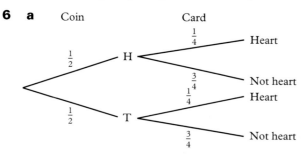

   b Probability of getting head and a heart $= \frac{1}{2} \times \frac{1}{4} = \frac{1}{8}$

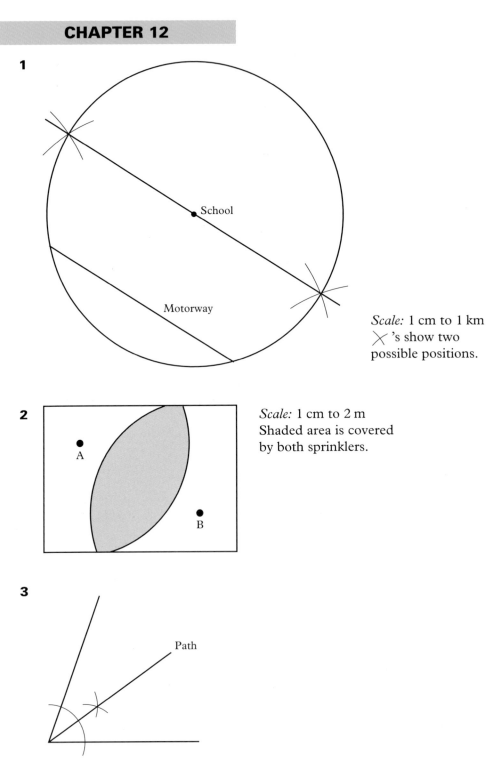

CHAPTER 12

1

School

Scale: 1 cm to 1 km
✕ 's show two
possible positions.

Motorway

2

A

B

Scale: 1 cm to 2 m
Shaded area is covered
by both sprinklers.

3

Path

4

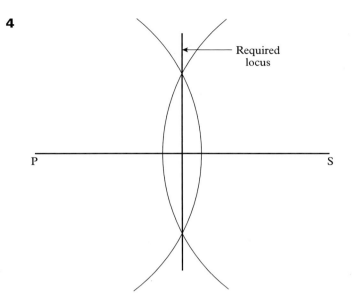

Required locus

P   S

CHAPTER 13

1 a $y = 5x - 2$ The largest number in front of the x is 5.
 b +5   The least steep line is $y = 2x + 5$ which has a y intercept of 5

2 $y = 2x - 4$ (3, 2) $2 \times 3 - 4 = 2$
   $y = x - 4$ (9, 5) $9 - 4 = 5$
   $y = 3x - 2$ (3, 7) $3 \times 3 - 2 = 7$

3 a $y = 2x$ The line is parallel to $y = 2x + 6$ so must have
                 $y = 2x$.
                 It passes through 0 so the equation is just $y = 2x$.
 b $y = -2x + 4$ The line is parallel to $y = -2x - 3$ so must have
                 $y = -2x$. It passes through 4 so the equation is
                 $y = -2x + 4$.

4 a (1) $2x + y = 11$
    (2) $5x - y = 17$
    Add (1) and (2) $7x \quad\;\; = 28$
                      $x = 4$
    Put $x = 4$ in (1) $8 + y = 11$
                      $y = 3$
    So $x = 4$, $y = 3$
    Check in (2) $5 \times 4 - 3 = 20 - 3 = 17$ ✓

b (1) $\qquad\qquad 7x + 2y = 31$
(2) $\qquad\qquad 3x + 5y = 34$
Multiply (1) by (5) and (2) by 2
$$35x + 10y = 155$$
$$6x + 10y = 68$$
Subtracting $\qquad 29x \qquad\quad = 87$
$$x = 3$$
Put $x = 3$ in (1) $\qquad 21 + 2y = 31$
$$2y = 10$$
$$y = 5$$
So $x = 3$, $y = 5$
Check in (2) $\quad 3 \times 3 + 5 \times 5 = 9 + 25 = 34$ ✓

5

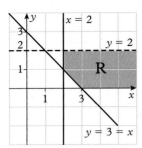

$x \geqslant 2$ shading needs to be on the right of solid line $x = 2$
$y < 2$ shading needs to be below dashed line $y = 2$
$y \geqslant 3 - x$ shading needs to be above solid line $y = 3 - x$

CHAPTER 14

1 a

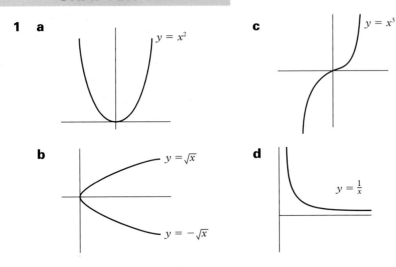

2 **a**

x	-4	-3	-2	-1	0	1	2	3
y	12	6	2	0	0	2	6	12

b

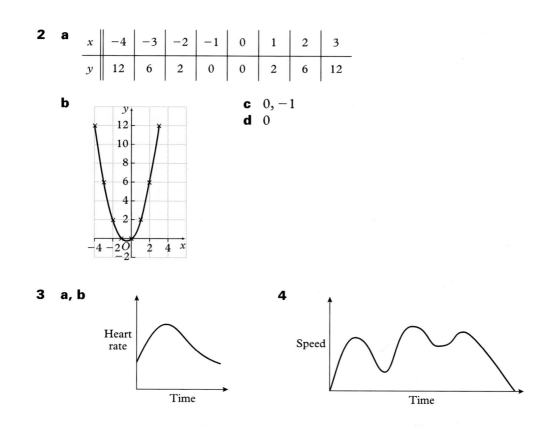

c $0, -1$

d 0

3 **a, b**

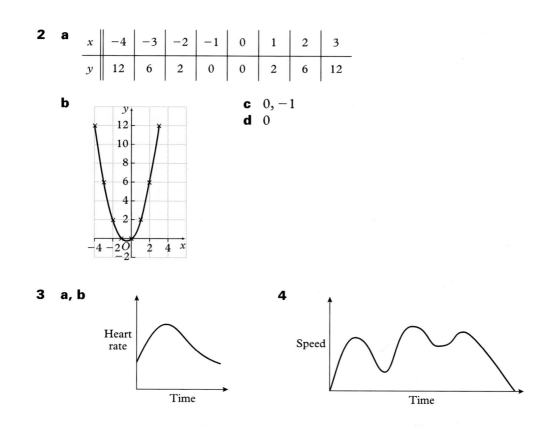

4

CHAPTER 15

1

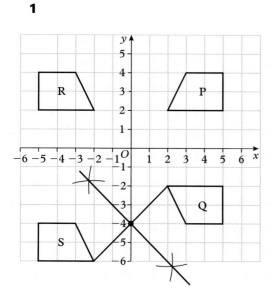

e To find the centre of rotation:
Join pairs of corresponding points
and construct the perpendicular
bisectors.
The crossing point of these is the
centre of rotation.
Centre of rotation is $(-4, 0)$
The angle of rotation is 180°.

f Rotation of 180°, centre of
rotation O.

2 b (1)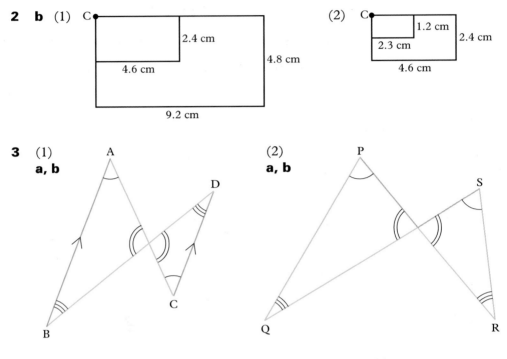

(2)

3 (1) **a, b**

(2) **a, b**

c Scale factor is $\dfrac{7.5}{5} = 1.5$

$w = 7 \times 1.5 = 10.5$ cm

$x = 5.4 \div 1.5 = 3.6$ cm

c Scale factor is $\dfrac{30}{24} = 1.25$

$y = 14 \times 1.25 = 17.5$ mm

$z = 35 \div 1.25 = 28$ mm

CHAPTER 16

1 $a =$ Total of interior angles $= 3 \times 180°$
$\qquad\qquad\qquad\qquad\qquad = 540°$

Total of 4 given angles $= 290°$

$a = 540° - 290° = 250°$

$b = 180° - 110° = 70°$

5

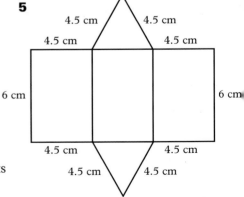

2 a 2   **c** 3
   b 1   **d** 2

3 A has 6 spots, B has 4 spots, C has 5 spots

4 a Pentagonal pyramid
   b Heptagonal prism

Exercise 1

1	480	**3**	8420	**5**	70 000
2	540	**4**	7770	**6**	90 030

Exercise 2

1	2700		**7**	415 300
2	9100		**8**	900 000
3	74 000		**9**	4 004 000
4	29 100		**10**	9 240 000
5	427 000		**11**	3 010 000
6	840 000		**12**	73 700 000

Exercise 3

1	780	**5**	1950	**9**	3550
2	840	**6**	1360	**10**	6080
3	780	**7**	3680	**11**	8730
4	690	**8**	1250	**12**	36 750

Exercise 4

1	74	**3**	189.1	**5**	208
2	325	**4**	723.4	**6**	4

Exercise 5

1	6586	**3**	1640	**5**	7039
2	2294	**4**	72 180	**6**	401

Exercise 6

1	21 460	**3**	5240	**5**	99 040
2	78 910	**4**	56 890	**6**	6010

Exercise 7

1	675	**5**	9550	**9**	11 990
2	1824	**6**	14 634	**10**	53 365
3	6519	**7**	17 595	**11**	53 534
4	10 504	**8**	26 628	**12**	88 021

Exercise 8

1	45	**5**	50	**9**	29
2	62	**6**	30	**10**	24
3	67	**7**	35	**11**	66
4	46	**8**	29	**12**	92

Exercise 9

1	$58\frac{1}{6}$		**7**	$15\frac{1}{7}$
2	$51\frac{1}{13}$		**8**	$30\frac{7}{17}$
3	$51\frac{1}{8}$		**9**	$16\frac{1}{13}$
4	$42\frac{1}{22}$		**10**	$17\frac{2}{9}$
5	$22\frac{11}{12}$		**11**	$202\frac{3}{7}$
6	$27\frac{1}{4}$		**12**	$581\frac{2}{5}$

Exercise 10

1	82		**5**	816
2	6		**6**	940
3	482		**7**	700
4	93		**8**	50 000

Exercise 11

1	54		**5**	840
2	71		**6**	84
3	82		**7**	400
4	64		**8**	40

Exercise 12

1	41	**5**	376
2	16	**6**	154
3	93	**7**	22
4	105	**8**	306

Exercise 13

1	3.27	**3**	0.34	**5**	0.301
2	9.64	**4**	0.879	**6**	1.03

Exercise 14

1	1.825	**3**	0.234	**5**	0.102
2	4.791	**4**	0.176	**6**	0.3102

Exercise 15

1	0.6914	**3**	0.0167	**5**	0.00671
2	0.8726	**4**	0.0342	**6**	0.00804

Exercise 16

1	$\frac{7}{9}$	**4**	$1\frac{3}{5}$
2	$\frac{11}{12}$	**5**	$1\frac{5}{8}$
3	$\frac{11}{12}$	**6**	$1\frac{6}{11}$

Exercise 17

1	$\frac{3}{8}$	**5**	$\frac{11}{18}$	**9**	$\frac{17}{30}$
2	$\frac{4}{15}$	**6**	$\frac{1}{2}$	**10**	$\frac{9}{14}$
3	$\frac{1}{2}$	**7**	$\frac{7}{12}$	**11**	$\frac{29}{40}$
4	$\frac{8}{9}$	**8**	$\frac{13}{21}$	**12**	$1\frac{1}{12}$

Exercise 18

1	$\frac{3}{8}$	**5**	$\frac{1}{2}$	**9**	$\frac{5}{24}$
2	$\frac{1}{5}$	**6**	$\frac{7}{12}$	**10**	$\frac{5}{12}$
3	$\frac{4}{11}$	**7**	$\frac{1}{12}$	**11**	$\frac{7}{24}$
4	$\frac{3}{10}$	**8**	$\frac{1}{6}$	**12**	$\frac{7}{44}$

Exercise 19

1	$\frac{1}{3}$	**5**	$\frac{2}{3}$	**9**	$\frac{5}{7}$
2	$\frac{1}{3}$	**6**	$\frac{3}{5}$	**10**	$\frac{1}{5}$
3	$\frac{1}{3}$	**7**	$\frac{2}{3}$	**11**	$\frac{1}{2}$
4	$\frac{2}{3}$	**8**	$\frac{2}{3}$	**12**	$\frac{19}{40}$

Exercise 20

1
- **a** 34 mm
- **b** 128 mm
- **c** 1310 mm
- **d** 1137 mm

2
- **a** 410 cm
- **b** 280 cm
- **c** 1210 cm
- **d** 32 400 cm

3
- **a** 8800 m
- **b** 9700 m
- **c** 15 000 m
- **d** 100 000 m

Exercise 21

1
- **a** 3000 g
- **b** 7000 g
- **c** 21 000 g
- **d** 3 kg
- **e** 6 kg
- **f** 0.8 kg

2
- **a** 3500 g
- **b** 7200 g
- **c** 3840 g
- **d** 6.5 kg
- **e** 3.2 kg
- **f** 2.8 kg

3
- **a** 5000 mg
- **b** 8 g
- **c** 1.64 g
- **d** 3 t
- **e** 4000 kg
- **f** 2500 kg